OUT OF CHINA

OUT OF CHINA

Responsible Port Innovation along the 21st Century Maritime Silk Road

Edited by

Wim Ravesteijn
Delft University of Technology, The Netherlands
and
Beijing Information Science and Technology University, China

WIT*PRESS* **Southampton, Boston**

Edited by:

Wim Ravesteijn
Delft University of Technology, The Netherlands
and
Beijing Information Science and Technology University, China

Editorial support provided by:

Zhijie Li
Delft University of Technology, The Netherlands and *Harbin Institute of Technology, China*

Yan Zhang
Delft University of Technology, The Netherlands and *Harbin Normal University, China*

Published by

WIT Press
Ashurst Lodge, Ashurst, Southampton, SO40 7AA, UK
Tel: 44 (0) 238 029 3223; Fax: 44 (0) 238 029 2853
E-Mail: witpress@witpress.com
http://www.witpress.com

For USA, Canada and Mexico

Computational Mechanics International Inc
25 Bridge Street, Billerica, MA 01821, USA
Tel: 978 667 5841; Fax: 978 667 7582
E-Mail: infousa@witpress.com
http://www.witpress.com

British Library Cataloguing-in-Publication Data

A Catalogue record for this book is available from the British Library

ISBN: 978-1-78466-453-4
eISBN: 978-1-78466-454-1

Library of Congress Catalog Card Number: 2021940103

CONTENTS

PREFACE

The fourth Needham question

Wim Ravesteijn

Department of Values, Technology and Innovation, Delft University of Technology, The Netherlands
Beijing Information Science and Technology University, China

How did China manage to catch up with and even overtake the West? This book might address one of the key points of the answer to this question, which will most probably occupy future historians when they look back in time from the year 2050 or so. It does not go to the roots of the present emergence of China, which have to do with the developments during the Mao era [1] and the subsequent Deng Xiao Peng era [2]. This volume, however, addresses what, in retrospect, in all likelihood will be seen as at least a leap forward, resulting from China's turn towards a much more active foreign policy. This is happening, as we speak, under the heading of the One Belt, One Road (OBOR) initiative or, in short, the Belt and Road Initiative (BRI).

This question could be spoken of as the 'fourth Needham question'. The work of Joseph Needham always impressed me, and during my courses on the History of Technology I liked to address the questions that inspired this famous Sinologist in his drive to uncover and explain China's great history in science, technology and medicine, work which is now, after his death in 1995, being continued by the Needham Research Institute in Cambridge [3]. He was driven by two questions:

- Why did science and technology undergo intense development in Europe during the last 500 years, whereas the growth of science and technology in China remained behind in that same period? (The first Needham question).
- Why did science, technology and industry undergo strong development in China at the time of the European Middle Ages? (The second Needham question).

There is an additional question (see e.g. [4]), attributed to the well-known sociologist Max Weber, which I also liked to discuss, and which was relevant at the TU Delft:

- Why did the Industrial Revolution take place in Europe and not in China? (The third Needham question or the Weber question).

These three questions remain interesting and relevant, but in this volume, I would like to look at the future, from the potential of the present. The BRI is an ambitious programme to reanimate the historical Silk Road [5], which might give China's economy a boost, simultaneously stimulating the development of other Asian countries and world trade and growth as a whole [6]. Consequently, it might enable China to recover its central place in world affairs, as suggested by the fourth Needham question.

The BRI also functions as a platform for the exchange of ideas and information, and this book and the underlying five-year research project reflect a similar process of knowledge sharing and mutual learning. The input of various visiting scholars from China, mostly PhD students and mostly supported by CSC scholarships, has been crucial. It was a great pleasure to work with these curious and diligent students and I am grateful for their contributions to this book. Conversely, I have not only enjoyed our cooperation but also greatly profited from it, hopefully, all for the better of this publication. For the same last five years or so, my mission has been to create opportunities for young, promising students, and I hope their Delft experience will help them in successfully continuing their academic careers.

This book concludes my professional career at TU Delft, where my activities played in the broad domain of technology and culture. Though it has not always been possible to occupy myself with the broad questions of history, relating to the rise of the West and the relative lagging behind of the East, these questions have always been central in my interest. I consider it a privilege to be able in this book to touch upon these broad questions, though in a very specific way by focusing on port development along the new Maritime Silk Road. I have always considered it a challenge to study the minute from a broad framework, to deal with the concrete in an abstract way, and this book is my latest attempt to do so.

So, for somebody with a broad historical interest, the Silk Road is a feast to get one's teeth into. There is even an Italian connection, in the person of Marco Polo, which I appreciate a lot, as part of the preparations for an academic career took place in Italy, in Sicily to be precise, where I conducted anthropological fieldwork into the questions of development and underdevelopment within Europe [7]. My exotic training as an anthropologist later brought me to the East, first Indonesia, with a trilogy about public works as the main result [8–10] and after that China, where I became a Visiting Professor at HIT and BISTU. For me, the Silk Road brings it all together, and this book is the materialization of that!

Some of my colleagues were prepared to write chapters, doing so beside and despite their busy work, and I appreciate that a lot. A number of promising MSc students also produced chapters for this book, and I respect their devotion and creativity in working with the concept of Responsible Innovation. Jia He and Yi Liu also played major parts in the research project, and I am happy they did. My colleagues in the Section of Ethics and Philosophy of Technology – part of the Values, Technology and Innovation Department at the Technology, Policy and Management Faculty and under the responsible leadership of especially Jeroen van den Hoven, Ibo van de Poel and Sabine Roeser – constituted the stimulating intellectual environment of my work, and I hope this book clarifies my role and position in the group. Thanks to all of them, I look back on my last major research and book project at TU Delft with pleasure and with a bit of regret that our common journey along the Silk Road ends here.

References

[1] Wertheim, W.F., *Evolution and Revolution: The Rising Waves of Emancipation*. London: Pelican, 1974.

[2] Chuntao, X., *China through the Ages: From Confucius to Deng*. Beijing: New World Press, 2009 (2 volumes).

[3] Needham, J., *Science and Civilisation in China*. Cambridge: Cambridge University Press, 1954–2004 (7 volumes).

[4] Yifu Lin, J., The Needham puzzle, the Weber question, and China's miracle: Long-term performance since the Sung dynasty. Paper prepared for the 'World Economic Performance: Past, Present and Future – Long Term Performance and Prospects of Australia and Major Asian Economies', Seminar on the Occasion of Angus Maddison's 80th Birthday, 5–6 December 2006, School of Economics, University of Queensland. https://www.tandfonline.com/doi/abs/10.1080/17538960701565053?needAccess=true&. Accessed on: 19 May 2021.

[5] Frankopan, P., *The Silk Roads: A New History of the World*. London: Bloomsbury Publishing, 2015.

[6] Frankopan, P., *The New Silk Roads: The Present and Future of the World*. London: Bloomsbury Publishing, 2018.

[7] Ravesteijn, W., Milena Restudied: A Study of Economic and Political Change in a Sicilian Village, MA thesis, University of Amsterdam, 1979.

[8] Ravesteijn, W. & Kop, J., *For Profit and Prosperity: The Contribution Made by Dutch Engineers to Public Works in Indonesia, 1800–2000.* Zaltbommel and Leiden: Aprilis and KITLV Press, 2008.

[9] Kop, J., Ravesteijn, W. & Kop, K., *Irrigation Revisited: An Anthology of Indonesian–Dutch Cooperation 1965–2014.* Delft and Jakarta: Eburon, 2015.

[10] Ravesteijn, W., *Engineering the Dutch Empire: Irrigation, the Colonial State and Ideology in Java 1832–1942.* Delft: Eburon, 2018.

INTRODUCTION

The challenge of a sustainable and responsible 21st Century Maritime Silk Road

Wim Ravesteijn[1,2] Zhijie Li[1,3] & Yan Zhang[1,4]

[1]Department of Values, Technology and Innovation, Delft University of Technology, The Netherlands
[2]Beijing Information Science and Technology University, China
[3]School of Management, Harbin Institute of Technology, China
[4]School of Management, Harbin Normal University, China

Abstract

This chapter introduces this volume by discussing the Old Silk Road and the current initiative of the Chinese government to revive this network of routes through the One Belt, One Road agenda of infrastructure development. It shows that this ambitious programme can have major consequences for people and the environment, both positive and negative. The reception of the New Silk Road plans, both in popular media and in the scientific world, is discussed, while the four lead questions and an overview of the parts and chapters are given.

Keywords: Four lead questions, heartland, Belt and Road Initiative, reception, impact, overview

I.1 Introduction

The deep-seaport of the future might very well work on electricity, generated in a sustainable way. In Rotterdam, the biggest port of Europe, APM Terminals runs ahead of this future: it has electrified its container handling operations in the Maasvlakte 2 port [1], while this new port itself, built in the North Sea, has made preparations for onshore electricity [2]. The latter has not been followed by specific actions, though onshore electricity has been installed for cruise ships moored close to the city centre [3]. However, on the other side of the Eurasian continent, in Chinese seaports, onshore electricity is seriously worked on, for example, in the ports of Dalian and Shanghai [4, 5].

The use of electricity might fit in with the construction and expansion of ports sustainably and responsibly if our aims are to keep the earth liveable and enable the world population to increasingly enjoy a 'common wealth' [6].

'Might' because it depends on the way electricity is generated. Most of it in Rotterdam, for instance, comes from coal-fired power stations! It also depends on other factors, like public acceptance. However, most seaports in the world need to change, not only to meet rising transportation and welfare demands but also to be part of the necessary transition to a sustainable and responsible world.

Large deep-seaports, together with their industrial activities, are hubs in international trade and, as such, they are the main motors of economic growth and socio-economic development in general, and, consequently, they are major emitters of CO_2 and other pollutants, while port construction itself might also entail ecological and social damage. Simultaneously, in a situation in which people worldwide are increasingly educated and informed, ports are more and more under public pressure to fight and prevent ecological and social damage in their operations and development plans. As a result, the search for a sustainable and responsible port has begun, in Europe [7] and elsewhere, for example, in China, where eco-ports have been introduced [8–10].

These ecological and social issues are most urgent in relatively poor areas, which undergo or are expected to undergo a transition from a poor and stagnated economy towards a rich, welfare society, based on trade and industry.

In the future, the three world-ports mentioned could be connected through a northern route [11], but nowadays trade goes through southern routes, part and parcel of the shipping network of the 'monsoon area,' consisting of the Arabic Sea and the Indian Ocean up to the South China Sea, as well as all the coast lands involved [12]. It is a region of great activity and expectations. The game-changer here is the Chinese One Belt, One Road (OBOR) initiative, which aims to revive the ancient Silk Road, both over land (the Belt) and across the sea (the Road) [13] (see Figure I.1). China itself is booming, and the OBOR initiative, or the Belt and Road Initiative (BRI), is meant to support and export this growth.

This volume explores the construction and expansion of large deep-seaports (or 'mainports') along the 21st Century Maritime Silk Road; it assesses and evaluates these activities on the basis of the OBOR policy goals and sustainability and responsibility demands and it gives suggestions for improvement. In addition, this book explores the concept of Responsible Innovation in relation to Sustainable Development and its contents and significance, especially in relation to port innovation.

The contributions have been made in the framework of a five-year research project on the subject, the lead questions of which were:

1. Are the port construction and expansion projects along the new Maritime Silk Road, both in China and elsewhere with Chinese investments and assistance, responsible and sustainable? And if not, how could activities be improved?

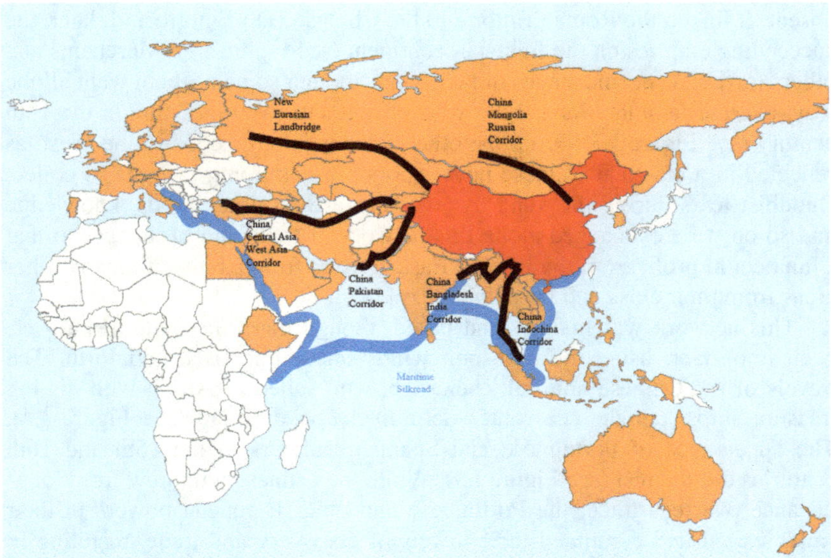

Figure I.1: The 21st Century One Belt, One Road vision. (Source: Wikimedia Commons, 2018 [14].)

2. Are these projects promoting socio-economic development, especially in the 'host countries' concerned? And if not, how could that be achieved?
3. Is China, as the initiator of the 21st Century Silk Road programme and in line with its intentions as suggested in all policy documents, showing 'moral leadership' in responsible and sustainable port innovation and development?
4. Together with Sustainable Development, Responsible Innovation constitutes the perspective from which we look at things, but what can we learn from port practices and related developments in terms of this concept? In other words, what would Responsible Port Innovation look like?

Questions 1, 2 and 3 will be more or less addressed in the various chapters. The final question will be addressed in the conclusion, with summarizing answers to the other questions as well. First of all, however, we will explore and specify this Chinese initiative and discuss the issues that we are facing here.

I.2 The Silk Road

I.2.1 Heartland

The Silk Road, in fact, a network of land and sea routes, connected East and West around 2000 years ago, though at times the contacts were meagre if not absent. It linked the Roman Empire to the Chinese Han Empire and, later, the succeeding empires on the Eurasian continent (see Figure I.2). Merchants and others travelled specific stages of the routes, though some of them went all the way, like the famous Marco Polo who visited and lived in China in the 13th century (see Figure I.3). From the other side, people travelled to the West, as reflected in a classical Chinese novel about a monk going to India to collect Buddhist texts during the Tang dynasty. For a long time, goods, knowledge and so on were exchanged along this network of trade and exchange, so that China could profit from the Middle East, Europe from China and many other areas from their close and far neighbours [15].

This network was mainly land based, though maritime trade has always been important, using the monsoon winds for sailing back and forth. The travels of the Chinese admiral Zheng He, who sailed to Africa with his big treasure ships, put the sea routes definitively on the map (see Figure I.4). The appearance of Portuguese and Spanish seafarers in the 15th and 16th centuries did the rest (see Figure I.5). While the Chinese withdrew from long-distance overseas trade, the Portuguese and other European powers in their wake started and continued their travels of discovery and trade, resulting in the European expansion and domination of the globe (until two world wars ended European hegemony in the 20th century). Consequently, maritime trade became predominant, affecting and displacing the earlier camel caravan trade. And that is still true: in our days, oil tankers and container ships largely determine world trade, with the monsoon area as the scene of the greater part of this trade; some one million ships cross the waters between Zanzibar (East Africa) and Singapore at the Strait of Malacca (between Malaysia and Sumatra, Indonesia) [12].

Nowadays, in retrospect, the historical Silk Road is increasingly seen as the motor of trade and development through the ages, its territories considered as the old 'Heartland' of world history [15, 16]. Cities and empires thrived in this area, although conquerors and other influences could cause temporary setbacks. Western colonial and imperial powers could control the Silk Road for their own interests, but the tables turned with the rise of strong states along the road, including India and China. The Chinese OBOR initiative aims to revitalize this network of land and sea routes.

I.2.2 The One Belt, One Road initiative

The OBOR initiative was launched in 2013 and it aspires to improve connectivity through the construction and improvement of ports, roads and pipelines across Asia, connecting China with Africa and Europe. The massive programme of investments and assistance aspires to promote economic growth as well as stability and peace in the whole region, while sustainability

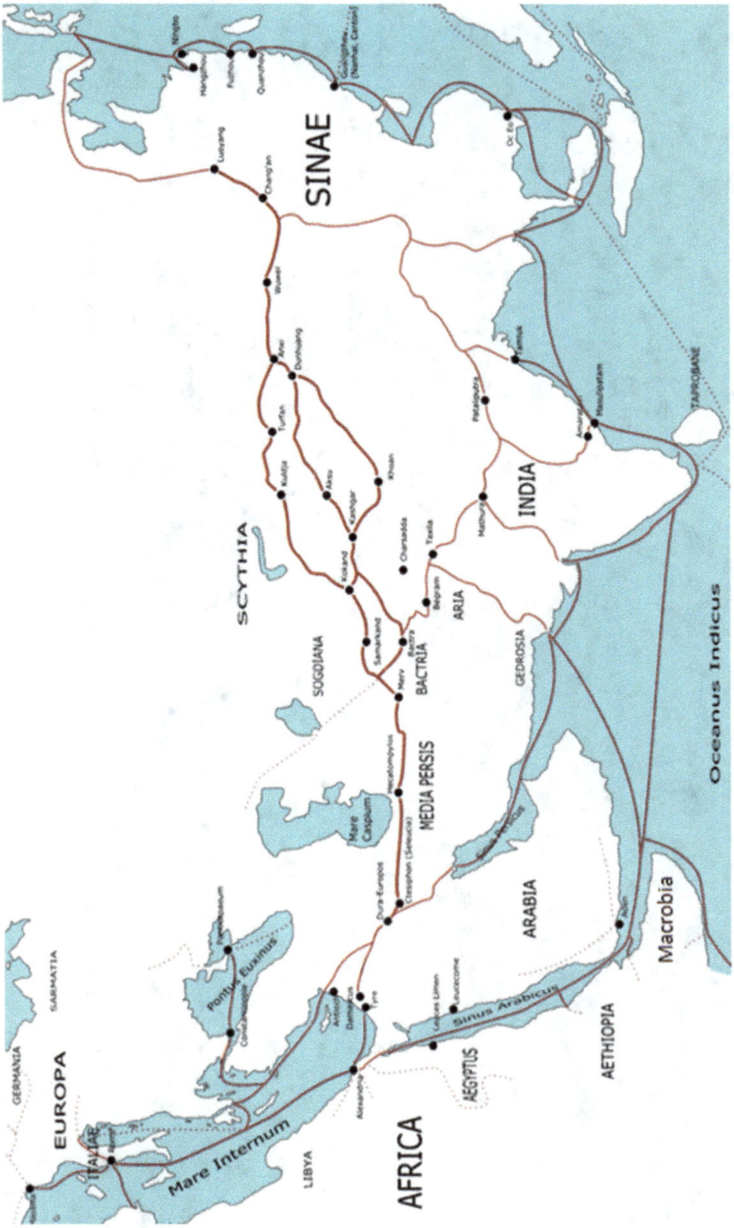

Figure I.2: The ancient Silk Road. (Source: Wikimedia Commons, 2018 [17].)

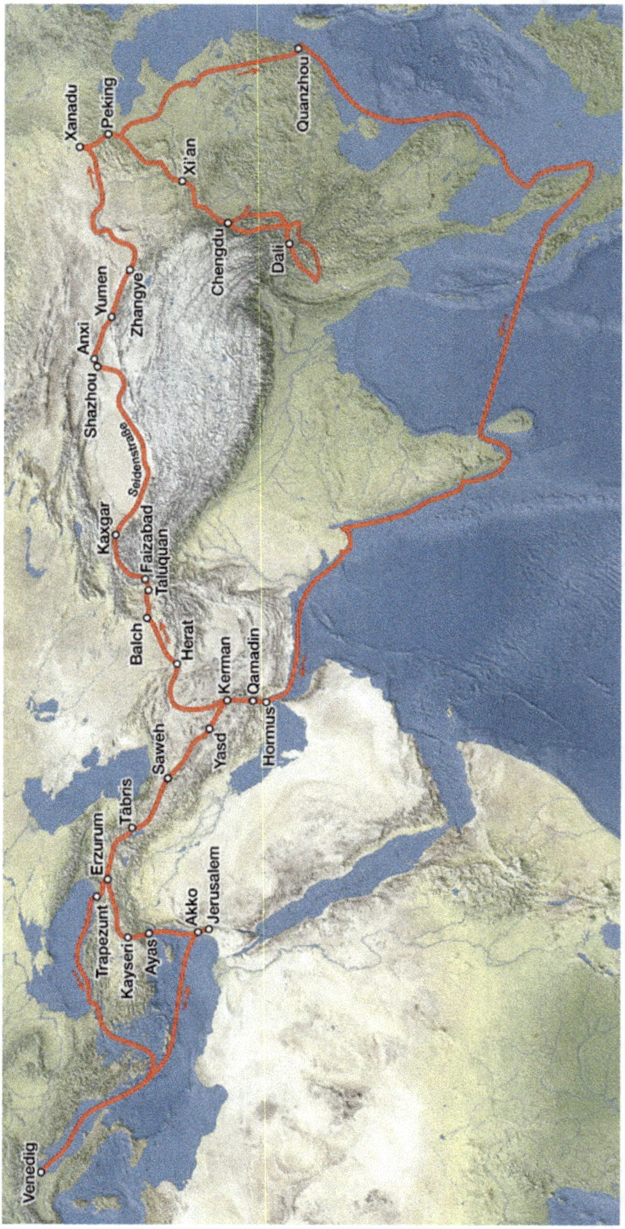

Figure I.3: The travels of Marco Polo. (Source: Wikimedia Commons, 2018 [18].)

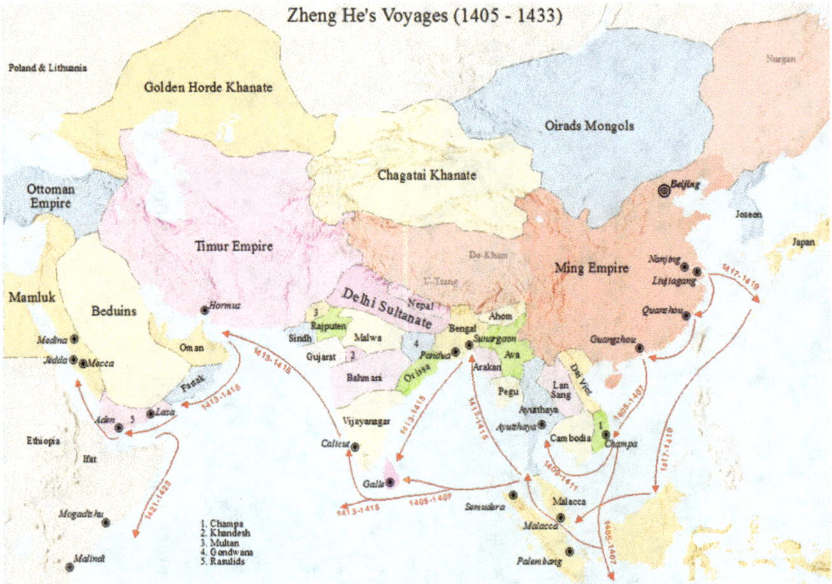

Figure I.4: The travels of Zheng He. (Source: Wikimedia Commons, 2018 [19].)

and cultural diversity are indicated to be important too [13, 21, 22]. The initiative follows upon the earlier 'String of Pearls' policy and it goes together with the revived (and cultivated) interest in the travels of Zheng He.

The main drivers behind the programme relate to China's own growing if not booming economy and its aspirations to improve prosperity throughout the country and, more basically, to its need for raw materials, on the one hand, especially oil and natural gas, which it imports from the Middle East and some other countries in the region, and, on the other hand, its need for foreign markets. However, the goals are broader. Under the OBOR initiative, China proposes and strives after maritime cooperation with other countries on the basis of principles like bridging differences and building consensus, openness, independence and inclusive development, market-based operation and multi-stakeholder participation as well as joint development and sharing benefits. With its openness, inclusiveness and win–win character, the initiative has attracted more and more countries to become involved. Until the end of 2017, the number of OBOR cooperation agreements had reached 100, with 86 countries or international organizations as participants. The Silk Road initiative

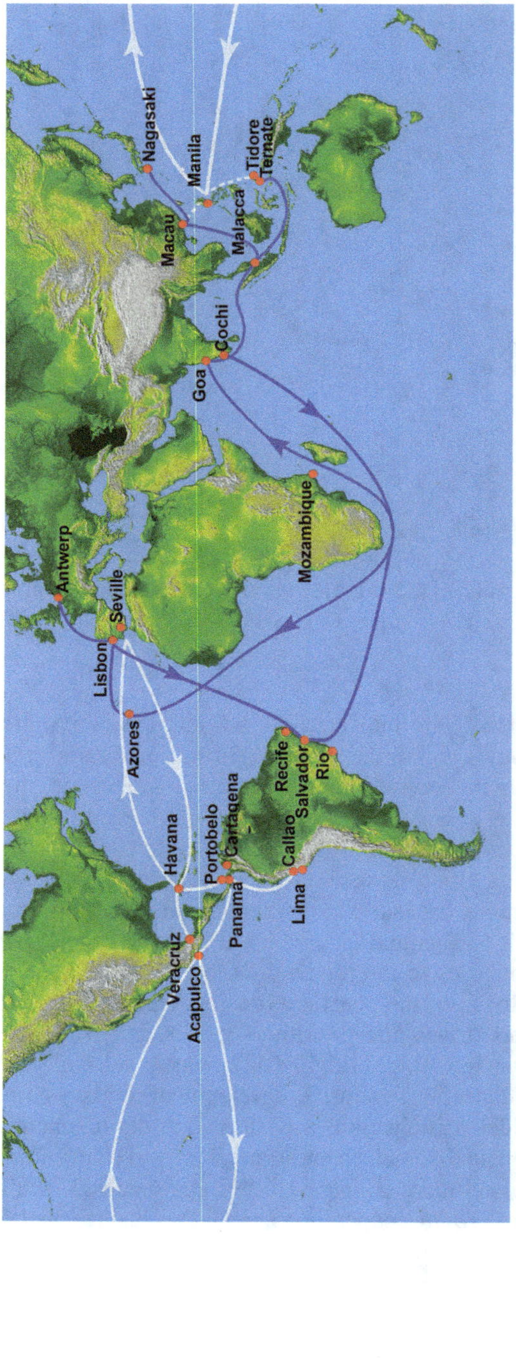

Figure I.5: Portuguese and Spanish explorations in the 16th century. (Source: Wikimedia Commons, 2018 [20].)

Table I.1: Foreign direct investment within the OBOR initiative.

Year	Number of countries involved	Main countries involved	Total investment amount
2015	49	Singapore, Kazakhstan, Laos, Indonesia, Russia, Thailand	US$14.82 billion
2016	53	Singapore, Indonesia, India, Thailand, Malaysia	US$14.53 billion
2017	59	Singapore, Malaysia, Laos, Indonesia, Kazakhstan, Vietnam, Russia, UAE, Cambodia	US$14.36 billion
2018 (Jan–Aug)	55	Singapore, Laos, Malaysia, Pakistan, Vietnam, Indonesia, Thailand, Cambodia	US$9.58 billion

is part of the broader 'Two Oceans Strategy,' through which China aspires to expand its range of activities in both the Indian and Pacific Oceans.

The year 2018 is the fifth year since the launch of the initiative. During these five years, from September 2014 until August 2018, an increased number of investment projects have been set up, especially after the US$40 billion investment fund was established [24] by the Silk Road Fund on 14 December 2014. Foreign direct investments have been stable, as shown in Table I.1 (see also Figure I.6). Most of the projects focus on infrastructure construction and capacity cooperation, mainly as to ports, railways, highways, electric power plants, aviation and communications.

China's maritime transport service, which all the countries involved in OBOR can access, has made great progress. Chinese companies have participated in the construction and operation of 42 ports in 34 countries [25], including Piraeus in Greece, Hambantota in Sri Lanka, Gwadar in Pakistan and so on. All over the world, China's shipping lane connections included more than 200 countries and 600 main ports.

Constructing and upgrading deep-seaports from Dalian to Venice is part of the Maritime Silk Road programme, including both domestic and foreign ports. Until now (2018), investments in domestic port projects have amounted to 176.592 billion RMB (US$26.8 billion), 16.95% of the whole OBOR budget, ranking second only to railway construction investment (see Table I.2 and also Figure I.7). For foreign ports, another 5000 billion RMB (US$759.2 million) has been made available (see Table I.3 and also Figure I.8) [26, 27].

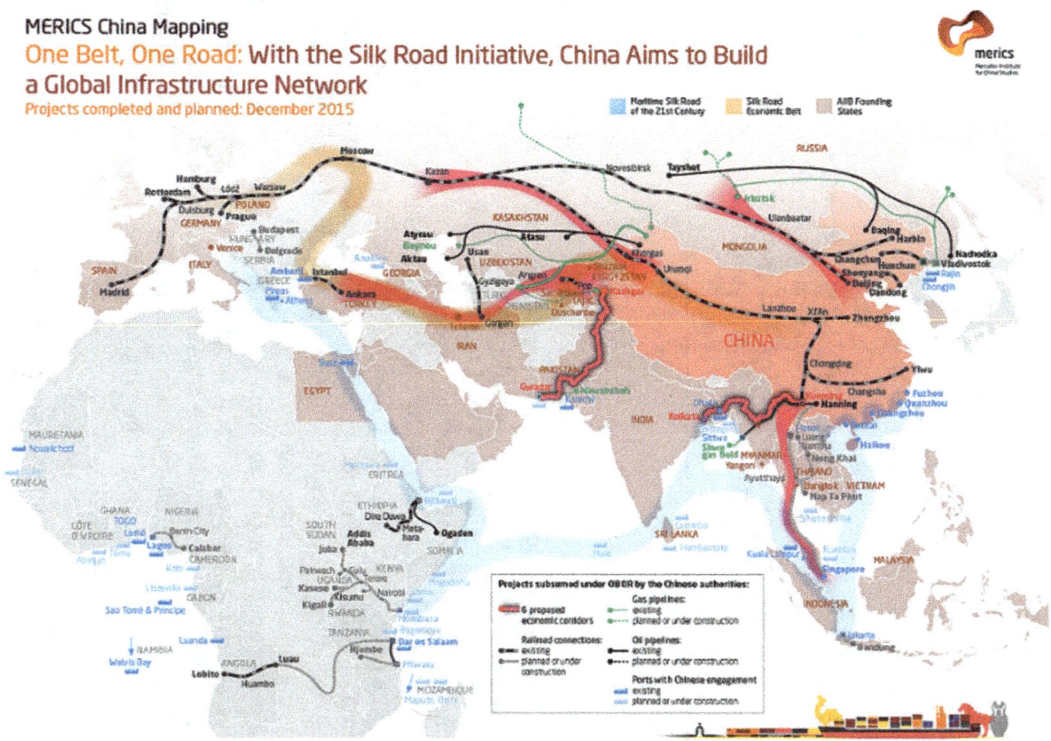

Figure I.6: Projects planned and completed under the OBOR initiative in December 2015. (Source: Merics, 2018 [23].)

Table I.2: Investments in Chinese ports.

Region	Port	New project investment (billion RMB)	Project process
Bohai region	Tianjin	1.843	Feb 2015
	Qingdao	7.269	Aug 2014 (approved)
	Dalian	0.451	Jan 2015
		3.77	Jan 2015 (approved)
	Yantai	0.54	Feb 2015
Yangtze Delta	Shanghai	1.285	Nov 2014 (approved)
	Ningbo-Zhoushan	0.787	Jul 2015
	Lianyun	0.625	May 2014 (operation)
Pearl River Delta	Guangzhou	5.552	Sept 2014
	Shenzhen	1.033	Nov 2014 (approved)
	Xiamen		Jan 2015 (approved)
	Fuzhou	0.67	
	Quanzhou	1	Apr 2014–Apr 2016
	Shantou		In construction
	Zhanjiang	0.969	Aug 2014 (approved)
	Haikou	2	In construction

I.2.3 Reception

The Belt and Road Initiative has been received with mixed feelings. Many countries have reacted positively, hoping and expecting the initiative will bring improvements to their countries, though this applies more to governments than opposition forces. Media attention has been overwhelming, with articles and views expressing both positive and negative aspects. Scientific papers and other studies generally reflect this picture.

The following positive points can be distilled from all the comments:

- The vision of a global economy has been a key and hot issue for the Chinese government and Chinese corporations, mostly focused on market participation and economic benefits. BRI projects and investments create opportunities and possibilities for profit and growth. Many countries have been involved in this grand strategy and, consequently, will benefit from official cooperation [30].
- In view of the proposal and practice of long-term projects in the BRI, China is really growing into a powerful and responsible major player on

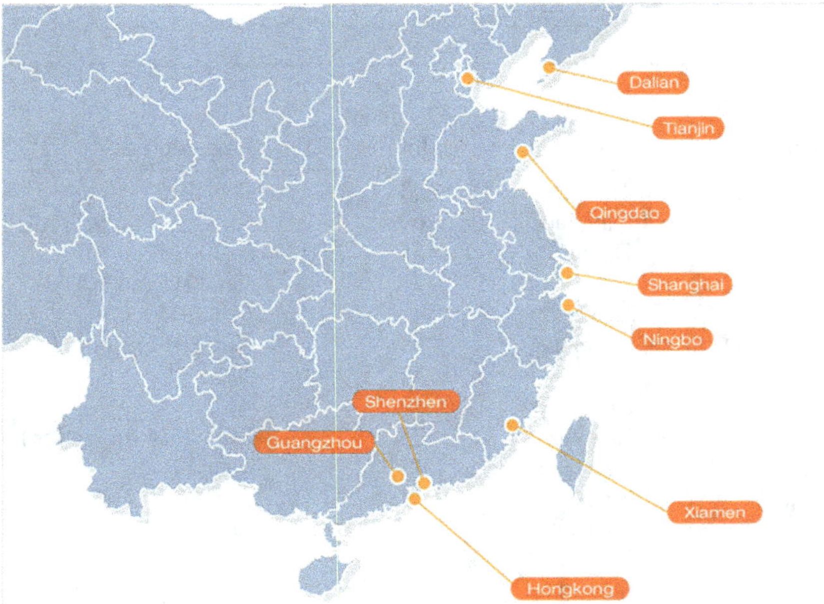

Figure I.7: Major Chinese ports. (Source: EC4U LTD, 2018 [28].)

the international stage [31], pursuing a global position and contributing to regional security [32], world trade, international financial-economic decision-making and so on [33]. All influence on other countries and international organizations has to be seen as 'soft power' characterizing the new era under the administration of Xi Jinping, without aggressive intentions, offensive actions or other forms of domination towards other countries, especially when it comes to underdeveloped and weak countries [34].

- The BRI is a specific strategy for regional development and cooperation, especially for the Asia-Pacific area as well as the Eurasian context [31, 35, 36]. Huge resources are being deployed in the domains of infrastructure and energy, besides investments and trade [37], not only in China but also in other Asian, African and European countries [38, 39].
- Many large road, railway and pipeline construction projects have been set up in 'host countries,' in which Chinese companies and governments cooperate with local governments and partners [39]. Some of them have already been implemented with works functioning according to plan, for example in African and Asian countries such as Turkey, Malaysia, Kenya, Egypt and so on [39, 40].

Table I.3: Investments in foreign ports.

Port	Chinese investing company	Amount	Starting
Piraeus	COSCO	0.3685 billion EUR	Apr 2016
Colombo	China Communications Construction Company Ltd.	US$1.396 billion (1.15 billion + 0.246 billion)	Sept 2014 + Mar 2022
Darwin	Landbridge Group	0.506 billion AUD	
Gwadar	COPHC	US$1.62 billion	
Istanbul Kumport	COSCO, China Merchants, Bank China Investment Corporation	US$0.9 billion	

- Several port investment plans and partnerships have been proposed involving the governments of China and other countries along the Maritime Silk Road [40–43].

However, it is also clear that the initiative has not remained without criticism. Issues raised include the following:

- The initiative requires the support of host country governments, which are often divided, apart from opposition forces. Politics in these countries

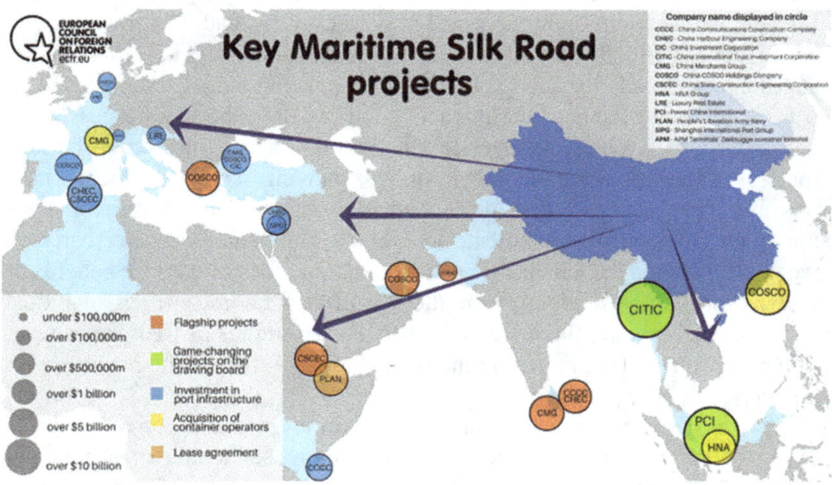

Figure I.8: Key Maritime Silk Road projects. (Source: Duchâtel & Sheldon Duplaix, 2018 [29].)

is often characterized by corruption, with states, at best, being 'soft,' i.e. inefficient. Large investments, without any demands, could aggravate these corruption and soft state problems [44].

- Global geopolitical aspects: China is supposed to have a hidden strategic agenda of domination and control in the region, instead of or beside all the expressed good intentions [35], [45–48].
- The plans are expected to have serious environmental impacts, which has received support from scientific studies [49–51].
- Participation in BRI projects can increase the foreign debts of host countries and, consequently, increase the dependency on Chinese companies, the Chinese government and international banks, including the Asian Infrastructure Investment Bank (AIIB), initiated by China. The financial burden will also be one of the major problems during trade, regional economic cooperation and investments in infrastructure [37, 48].
- In view of the perceived costs and benefits, public acceptance in the countries involved is often deficient, as people are suspicious and afraid of being dominated by China as a new upcoming world power [46, 52].
- There are security problems in many involved countries related to terrorism, civil wars and failed states, which are not beneficial to public works construction [53].
- The investment policies of Chinese companies are not transparent, which, again, opens the door to suspicion and fear [52, 54].

All of these aspects are important for this study, though the geopolitical facts, which get ample attention in the world of politics, news and, to a lesser extent, science, are not directly discussed here. The discussion of the BRI in the news, scientific papers and otherwise shows some interesting biases and gaps.

First, the main focus of the impacts is on the Belt, the land routes, while the Road, the sea routes, receives less attention. Second, most contributions express serious doubts about the initiative, viewing it as a political threat or as a comprehensive geopolitical challenge, rather than a programme for economic growth and development. Also, many commentators and researchers have not fully taken the BRI into consideration, only focusing on the benefits or concerns for their own countries or, the scientists, for some special analytical purposes. Furthermore, there are no thorough case studies of BRI projects, let alone port projects, and no comprehensive studies of the BRI in general (with the exception of [55]).

This study might contribute to filling in these gaps and correcting these biases by emphasizing port projects from economic, social and environmental perspectives, calling attention to positive aspects, besides negative ones. With its studies of cases in and outside China, this volume gives both a detailed and, taken together, a comprehensive picture of the effects of the BRI for both

Chinese and regional and global development. In addition, technologically and in terms of policies and otherwise, possibilities are explored and suggestions advanced to strengthen positive aspects and avoid or mitigate the negative aspects. Leading in that respect is our perspective, as will be explained below. Before we do so, we will elaborate on the impact, as this is not only important from our Responsible Innovation and Sustainable Development approach, but it even justifies applying and developing this approach.

I.2.4 Desired and undesired consequences

The monsoon area is affected by large socio-economic inequalities. Some regions are flourishing and strongly developing, Gujarat for example. Other areas are struck by poverty and underdeveloped. Development and underdevelopment, modernization and (increasing) backwardness sometimes go together, like in Gwadar, Pakistan. The political picture also shows great diversity. There are democratic states (like India), authoritarian states (like Oman), there are states showing a mixed character (like Bangladesh) and there are failed states (like Afghanistan). In addition, the area is the scene of civil war and Muslim terrorism, besides rising middle classes engaged in trade [12]. The OBOR initiative may change all of this, hopefully supporting positive developments and countering negative ones.

The impact of this initiative might be enormous, not only economically but also socially and ecologically. To assess the impact, a full 'Human Development Impact assessment' study [56] would be necessary for the programme in general and all the specific projects it comprises. This includes economics (at the project level: profitability and employment; at the national level: growth and rising incomes), social aspects (equal income distribution and welfare), sustainability (ecological soundness) and empowerment of people (as stakeholders and citizens). More specifically, the following can be said about this.

The OBOR initiative might give the whole area an economic boost, resulting in higher incomes and increased prosperity, while in a number of backward regions it might induce political stability and enable the transition to a growth economy and welfare society. OBOR fits in with the development strategies of various countries because of shared basic orientations, one of which is economic growth.

China, as the second-largest economy in the world, has kept a high economic growth speed since 1978, the first year of the opening-up policy and also the onset of industrialization. In 2017, when the period of the highest growth figures is over, GDP still grew up to 6.9%, 2.3% higher than the US. In June 2017, the *Vision for Maritime Cooperation under the Belt and Road Initiative* document was issued by the National Development and Reform

Commission (NDRC) and the State Oceanic Administration (SOA) [57]. As mentioned in the Vision, steady progress had been achieved under high-level guidance and facilitation, while boosting the role of cooperation platforms, increasing financial investments and China's regions promoting the opening-up policy. The accomplishments until now show the world China's capabilities and determination to reach the goal of shared growth, while they can also inspire confidence in OBOR among the countries along the Silk Road. The development of the travel industry in Sri Lanka is a good example. In 2017, the number of tourists from China reached 268,000, while it was almost zero ten years earlier. In addition, the Maritime Silk Road is helping to upgrade its status to an international trading centre, with the Hambantota port as a (future) main hub. There will be another US$24 billion investment for Sri Lanka, added to the eight billion until now.

The OBOR initiative comes from China, but the opportunities and advantages it brings may benefit the whole world. The economic impact OBOR has already achieved testifies to this. At the Boao Forum for Asia Annual Conference 2018 held in Hainan, President Xi emphasized this. In addition, in relation to island economies, the Maritime Silk Road has been a topic of the Globalization and One Belt, One Road theme at the 21st Century Maritime Silk Road: Islands Economic Cooperation Forum [58]. Inclusive economic development has been highly stressed there. Consequently, we have every reason to believe OBOR will generate additional positive economic impacts in the near future and beyond.

The programme might also support a social transition to better living conditions and a welfare society, based on broad employment opportunities. With current population growth figures, employment stress is a great challenge. Big differences exist in the scale, age and structure of the labour force in the countries along the Road. In the past ten years, the average unemployment rate has been around 9% in the area, but it was 3.8% for ASEAN (Association of Southeast Asian Nations) countries and 13.1% for Central and Eastern Europe [59]. According to incomplete statistics of the Chinese Ministry of Commerce, the investments China made in the countries involved in OBOR have created about 180,000 employment opportunities in the host countries [60]. Hiring local employees and skill training and studying opportunities in China are effective measures that have been taken to improve the employment rate, promote employment ability and enhance labour quality.

The host countries hugely benefit from the infrastructure being built. The China Railway Express, also known as the Golden Silk Road, with stations in 28 Chinese cities and 29 European cities in 11 countries, has become the backbone of transportation in the OBOR area. According to the 2016 Report on Development of China's Outward Investment and Economic Cooperation, in 2015, of the total newly signed project contracts of Chinese companies with countries involved in OBOR, electric power projects covered 27.4%, transportation construction 16.2% and housing construction 15.7%. All these infrastructure building projects can boost economic development and benefit social life in host countries.

However, the ecological consequences might be serious. To sustain its economic growth, China alone might need the natural resources of another planet Earth in 15 years or so and, if successful in terms of economic growth across Asia as hoped and expected, the new connections will only increase the demand for raw materials. In addition, China is already experiencing great environmental problems in terms of harmful emissions and pollution in general, and the OBOR initiative could deteriorate ecological conditions elsewhere as well.

The need for Sustainable Development places high demands on technology, which may appear from the application of the IPAT formula: $I = P \times A \times T$, where a change in the ecological impact (I) is the product of population growth (P), improved affluence level (A) and resource-intensiveness of technology (T). From a sustainability perspective, generally speaking, the lower T is the more efficient and therefore more advanced the technology will be. To make a very rough estimate: if the population in the Silk Road countries increases by about 10% in the next decade or so and the standard of living rises by 100%, while we would like to reduce the ecological impact by 50%, technology should become more than four times more efficient, that is 400% [61]!

The gigantic scale of this task appears from a simple exercise: utilizing the IPAT variables to analyse the increasing ecological pressure in retrospect, which generates the results as shown in Figure I.9 and Table I.4. The variable I here represents CO_2 emissions, expressing human impact on the environment. The other three factors are the population (P), GDP per capita (A) and CO_2 emissions in relation to GDP (T). As shown in Table I.4, in three different phases between 1980 and 2015, CO_2 emissions decreased by 34%, 40% and 30%, respectively. In the year 2015 alone, CO_2 emissions related to GDP declined by 8%, which was not an easy goal to reach. In view of high population and GDP growth rates, environmental issues should be balanced by the development of clean energy technologies, supported by effective policies. Given the progress made until now, this means that the efforts have to be increased by a factor of ten. Consequently, as the biggest CO_2 emitter in the world, China still has a long way to go!

To control and even compensate for the environmental impact of OBOR and China's promotion of the Blue Economy [63], the countries involved should be aware of the significance of evolving new energy technologies. China has put a lot of effort into research and development regarding energy-related science and technology. Some innovations are even world leaders, for instance,

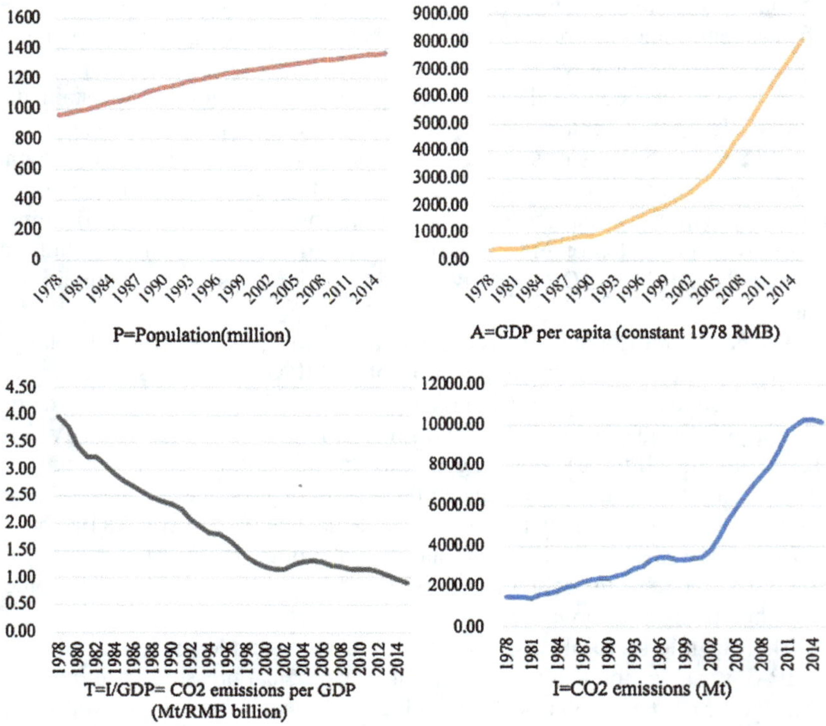

Figure I.9: China's IPAT performance since the opening-up policy. (Source: Global Carbon Atlas and China Statistical Yearbook [62].)

Table I.4: CO_2 emissions and three factors in China since 1980 (growth in %).

Period	I = CO_2 emissions	P = Population	A = GDP per capita (constant price)	T = CO_2 emissions per GDP
1980–1991	75%	17%	126%	−34%
1992–2003	69%	10%	155%	−40%
2004–2015	94%	6%	163%	−30%

in the domain of solar technology and equipment and when it comes to eco-cities [64].

Implementing the OBOR initiative in a sustainable and responsible manner is a huge socio-technical challenge. The development of advanced energy technologies and the realization of fruitful practices of clean and renewable energy use are necessary, as is intensive communication about achievements. Social benefits may follow economic ones, but 'common wealth,' all citizens sharing in the advantages, may only be attained if the right conditions are in place.

In order to maximize the benefits and minimize the costs, economically, socially and ecologically, the OBOR initiative should be realized sustainably and responsibly, which is provided for in the *Vision* and other documents, though lacking specifics. This volume aspires to fill in the gaps and give suggestions for how projects could be carried out with maximum benefit and minimum cost.

I.3 Overview of this volume

Practically, this volume aims to contribute to responsible and sustainable port innovation and development along the new Maritime Silk Road, while theoretically, it aims at getting a better understanding of what this means. The concepts of Responsible Innovation and Sustainable Development are both guiding and structuring principles and topics of reflection and consideration. Focused on and elaborated for ports, both concepts merge into a new concept – Responsible Port Innovation – and the ultimate theoretical goal of this book is therefore an exploration and clarification of this concept. The core of the book consists of a series of case studies of deep-seaports, large and leading world-ports or 'mainports.' The contributions are grouped into five parts.

Part I starts with presenting our research and evaluation framework, based on insights and standards related to technological innovation as well as economic growth and development. Consequently, things are put into a broader perspective by a contribution about the Asian Silk Road area as a whole, the challenges the countries involved face, their religious and cultural backgrounds and the social and cultural conditions needed to successfully implement Responsible Innovation and make it useful for socio-economic development in a country. Then attention is focused on China's foreign policy, of which the OBOR initiative forms an integral if not the key part. A chapter on eco-city development in China, a technological domain in which the country is quite successful, concludes this part. The subject of investigation here is what lessons can be drawn for 'eco-port' development or, in our terms, Responsible Port Innovation.

Port innovation and port-based development along the 21st Century Maritime Silk Road are discussed in Part II, in the form of several practical examples. Responsible Innovation and Sustainable Development provide the perspectives from which these cases are discussed, showing and evaluating what happens on the ground and how things could be improved. More cases follow in Part III. However, the focus here is on means and methods for expanding, completing and specifying Responsible Port Innovation.

Finally, the balance is drawn up on the basis of the four basic questions of this study. We will indicate how the 21st Century Maritime Silk Road can be made sustainable and responsible and how we can improve Responsible Port Innovation for this purpose.

References

[1] APM Terminals, https://www.apmterminals.com/en/news/news-releases/2017/news/first-commercial-vessel-call. Accessed on: 19 May 2021.

[2] Ravesteijn, W., Responsible port innovation in the Dutch delta: The Maasvlakte 2 port extension project. In: Ortt, J.R., Putten, D. van, Kamp, L.M. & Poel, I.R. van de (eds), *Responsible Innovation in Large Technological Systems*, Routledge, Abingdon, pp. 115–137, 2020.

[3] Lieshout, A., Viability Study of Sustainable Shore Side Electricity for Container Vessels in the Port of Rotterdam. Master's thesis, Delft University of Technology, 2019.

[4] Yan, P. & Ravesteijn, W., Chinese road to responsible innovation: Constructing a green port in Dalian. *WIT Transactions on The Built Environment*, 187, pp. 109–119, 2019.

[5] Li, Z. & Ravesteijn, W., this volume, Chapter 5.

[6] Sachs, J., *Common Wealth: Economics for a Crowded Planet*. Penguin, London, 2008.

[7] ESPO, Green Guide: Towards excellence in port environmental management and sustainability. European Sea Ports Organization (ESPO), Brussels, 2012.

[8] Tzimitras, V., Towards a Low Emission Port: Development of a decision making tool for technologies for the reduction of the shore-related emission footprint of existing ports in respect of stakeholders' values. Master's thesis, Delft University of Technology, 2018.

[9] Fifth Session, http://news.xinhuanet.com/politics/2005-10/11/content_3606215.htm, 2005. Accessed in 2013.

[10] Fifth Session, http://news.xinhuanet.com/misc/2006-03/16/content_4309517.htm, 2006. Accessed in 2013.

[11] Ravesteijn, W., Liu, Y. & Ping, Y., Responsible innovation in port development: The Rotterdam Maasvlakte 2 and the Dalian Dayao Bay

extension projects. *Water Science & Technology*, 72(5), pp. 665–677, 2015.

[12] Kaplan, R.D, *Monsoon: The Indian Ocean and the Future of American Power*. Random House, London, 2011.

[13] Belt and Road Forum for International Cooperation, Vision and actions on jointly building Belt and Road. http://2017.beltandroadforum.org/english/n100/2017/0410/c22-45.html. Accessed on: 2 July 2021.

[14] Wikimedia Commons, File:One-belt-one-road.svg. Accessed on: 7 November 2018.

[15] Frankopan, P., *The Silk Roads: A New History of the World*. Bloomsbury Publishing, London, 2015.

[16] China has a vastly ambitious plan to connect the world. *The Economist* 28 August 2018. Online: https://www.economist.com/briefing/2018/07/26/china-has-a-vastly-ambitious-plan-to-connect-the-world. Accessed on: 24 October 2018.

[17] Wikimedia Commons, File: Transasia trade routes 1stC CE gr2.png. Accessed on: 7 November 2018.

[18] Wikimedia Commons, File: Travels of Marco Polo.jpg. Accessed on: 7 November 2018.

[19] Wikimedia Commons, File: Voyages of Zheng He.png. Accessed on: 7 November 2018.

[20] Wikimedia Commons, File: 16th century Portuguese Spanish trade routes.png. Accessed on: 7 November 2018.

[21] Office of the Leading Group for the Belt and Road Initiative, Building the Belt and Road: Concept, Practice and China's Contribution, Beijing, 2017.

[22] Ministry of Ecology and Environment The People's Republic of China, The Belt and Road Ecological and Environmental Co-operation Plan. http://english.mee.gov.cn/Resources/Policies/policies/Frameworkp1/201706/t20170628_416869.shtml. Accessed on: 2 July 2021.

[23] Merics China mapping, Infographic/China Mapping Silk Road Initiative – Infographic/China Mapping – Publications – Mercator Institute for China Studies, https://chinahistory101.wordpress.com/2017/02/16/infographicchina-mapping-silk-road-initiative-infographicchina-mapping-publications-mercator-institute-for-china-studies/. Accessed on: 4 October 2018.

[24] Contribute to Belt and Road Initiative, Promote Connectivity, Silk Road Fund, http://www.silkroadfund.com.cn/enweb/23773/index.html. Accessed on: 5 June 2018.

[25] The fifth anniversary of the "Belt and Road" initiative, Chinese companies participated in the construction and operation of 42 ports in 34 countries. https://www.sohu.com/a/250253989_237762. Accessed on: 2 July 2021.

[26] Piraeus port becomes the pearl of one belt one road initiative, Xinhua News Agency, http://www.xinhuanet.com/world/2016-06/28/c_1119129171.htm. Accessed on: 20 August 2018.

[27] Project case 'One Belt One Road' PPP project case-Sri Lanka Colombo Port City Project. https://www.sohu.com/a/193477054_99922412. Accessed on: 2 July 2021.

[28] EC4U LTD, http://ec4ulimited.com/what-is-the-best-port-to-import-a-shipment-from-china-into-the-uk/. Accessed on: 4 October 2018.

[29] Duchâtel, M. & Sheldon Duplaix, A., Blue China: Navigating the Maritime Silk Road to Europe, Policy Brief, 23rd April 2018. https://www.ecfr.eu/publications/summary/blue_china_navigating_the_maritime_silk_road_to_europe. Accessed on: 4 October 2018.

[30] Swaine, M.D., Chinese views and commentary on the 'One Belt, One Road' initiative. *China Leadership Monitor*, 47(2), pp. 3–27, 2015.

[31] Ferdinand, P., Westward ho: The China dream and 'one belt, one road': Chinese foreign policy under Xi Jinping. *International Affairs*, 92(4), pp. 941–957, 2016.

[32] Clarke, M., 'One belt, one road' and China's emerging Afghanistan dilemma. *Australian Journal of International Affairs*, 70(5), pp. 563–579, 2016.

[33] Liu, W., Scientific understanding of the Belt and Road Initiative of China and related research themes. *Progress in Geography*, 34(5), pp. 538–544, 2015. In Chinese.

[34] Cheng, L.K., Three questions on China's 'Belt and Road Initiative.' *China Economic Review*, 40, pp. 309–313, 2016.

[35] Rolland, N., *China's Eurasian Century?: Political and Strategic Implications of the Belt and Road Initiative*. National Bureau of Asian Research, Seattle, WA, 2017.

[36] Zhao, M., The belt and road initiative and its implications for China—Europe relations. *The International Spectator*, 51(4), pp. 109–118, 2016.

[37] Dollar, D., China's rise as a regional and global power. *Horizons*, 2015(4), pp. 162–173, 2015.

[38] Van der Putten, F. P., *Chinese Investment in the Port of Piraeus, Greece: The Relevance for the EU and the Netherlands*, Clingendael Institute, The Hague, 2014.

[39] ZiroMwatela, R. & Zhao C., Africa in China's One Belt, One Road Initiative: A critical analysis. *ISOR Journal of Humanities and Social Sciences*, 21, pp. 21–34, 2016.

[40] Djankov, S. & Miner, S., *China's Belt and Road Initiative: Motives, Scope, and Challenges*. Peterson Institute for International Economics, Washington, DC, 2016.

[41] Michalakeas, D. & Garay Garcia, O. this volume, Chapter 6.

[42] Van Giezen et al., this volume, Chapter 7.

[43] Tzimitras, V., Towards a Low Emission Port: Development of a decision making tool for technologies for the reduction of the shore-related emission footprint of existing ports in respect to the stakeholders' values. Master's thesis, Delft University of Technology, 2017.

[44] Guo, R. & Yang, K., China's New Silk-Road Initiative and the South China Sea Disputes: Toward a New Mutual Deterrence Equilibrium. *Return of Geopolitics*, University of Arizona, Tucson, April, pp. 3–5, 2016.

[45] Sidaway, J.D. & Woon, C.Y., Chinese narratives on 'One Belt, One Road' (一带一路) in geopolitical and imperial contexts. *The Professional Geographer*, 69(4), pp. 591–603, 2017.

[46] Brewster, D., Silk roads and strings of pearls: The strategic geography of China's new pathways in the Indian Ocean. *Geopolitics*, 22(2), pp. 269–291, 2017.

[47] Zhao, Y., Introduction. In: Zhao, Y. (ed.), *International Governance and the Rule of Law in China under the Belt and Road Initiative*, Cambridge University Press, Cambridge, pp. 1–6, 2018. DOI: 10.1017/9781108332651.001.

[48] Szechenyi, N., (ed.), *China's Maritime Silk Road: Strategic and Economic Implications for the Indo-Pacific Region*. CSIS, Washington, DC, 2018.

[49] Laurence, B., China-backed Sumatran dam threatens the rarest ape in the world. *The Jakarta Post*, 6 May 2018. http://www.thejakartapost.com/academia/2018/05/06/china-backed-sumatran-dam-threatens-the-rarest-ape-in-the-world.html?fbclid=IwAR39-1Ir8fbwd5BBspPsn4sxKA47BR7BjQeJVk8-0rqmx5kvruw6qE1w3po. Accessed on: 28 November 2018.

[50] Tracy, E.F., Shvarts, E., Simonov, E. & Babenko, M., China's new Eurasian ambitions: The environmental risks of the Silk Road Economic Belt. *Eurasian Geography and Economics*, 58(1), pp. 56–88, 2017.

[51] Ascensão, F., Fahrig, L., Clevenger, A.P., Corlett, R.T., Jaeger, J.A.G., Laurance, W.F. & Pereira, H.M., Environmental challenges for the Belt and Road Initiative. *Nature Sustainability*, 1, May 2018, pp. 206–209. www.glel.carleton.ca/PDF/landPub/18/18AscensaoetalNatureSustainability.pdf. Accessed on: 4 June 2021.

[52] To rationally treat the negative judgement of Belt and Road Initiative, http://field.10jqka.com.cn/20180831/c606895193.shtml. Accessed on: 5 September 2018.

[53] China's belt-and-road plans are to be welcomed – and worried about. *The Economist*, https://www.economist.com/leaders/2018/07/26/chinas-belt-and-road-plans-are-to-be-welcomed-and-worried-about. Accessed on: 26 July 2018.

[54] Trump first shows discontent with the Belt and Road Initiative in public, https://www.sohu.com/a/246292241_825949. Accessed on: 2 September 2018.

[55] Frankopan, P., *The New Silk Roads: The Present and Future of the World*, Bloomsbury Publishing, London, 2018.

[56] Adisti, R., Human Development Impact of Broadband in Indonesia. Master's thesis, Delft University of Technology, 2017.

[57] National Development and Reform Commission (NDRC) & the State Oceanic Administration (SOA), Vision for Maritime Cooperation under the Belt and Road Initiative, 2017. http://www.xinhuanet.com/english/2017-06/20/c_136380414.htm. Accessed on: 29 November 2018.

[58] Session 26 21st Century Maritime Silk Road: Islands Economic Cooperation Forum. https://web.archive.org/web/20170828134017/http://english.boaoforum.org/2017flt25en/30946.jhtml. Accessed on: 2 July 2021.

[59] Zhang, Y. & Liu, L., Labour markets comparision and enlightenment in the nations along the Belt and Road Initiative. *West Forum* 27, 2017.

[60] China's direct investment in countries along the "Belt and Road" increased 70 times in 13 years. https://www.sohu.com/a/139927360_123753. Accessed on: 2 July 2021.

[61] Mulder, K, *Sustainable Development for Engineers: A Handbook and Resource Guide*. Greenleaf Publishing Ltd, Sheffield, 2006.

[62] Global Carbon Atlas, http://www.globalcarbonatlas.org/en/content/welcome-carbon-atlas. Accessed on: 19 May 2021; and China Statistical Yearbook. Beijing: China Statistics Press, 1982–present.

[63] Zhang, Y. & Ravesteijn, W., Sustainable port development based on the Blue Economy Framework: The example of Qingdao port. *WIT Transactions on The Built Environment*, 187, pp. 121–131, 2019.

[64] Lu, H. et al., this volume, Chapter 4.

Part I

Approach and context

CHAPTER 1

Research approach: Innovation, development and leadership

Wim Ravesteijn

Department of Values, Technology and Innovation, Delft University of Technology, The Netherlands
Beijing Information Science and Technology University, China

Abstract

This chapter presents the research and evaluation framework that underlies this volume. It is based on insights and standards related to technological innovation and economic growth and development. Responsible Innovation and Sustainable Development are lead concepts.

Keywords: Responsible Innovation, economic growth and development, Chinese model, Western model, role of technology, peace and prosperity, Sustainable Development, diffusion, circulation, leadership, research questions

1.1 Introduction

The relationship between technology and development is complex and part of a broader debate on development strategies in which economists and other scholars draw different lessons from the history of the West, the post-war experience with development aid and cooperation as well as the rise of emerging economies, especially in Asia (first, the 'Asian tigers,' later India and China). First, a new technological innovation should be embedded in society, which requires at least the acceptance of the people concerned, though meeting their values through Responsible Innovation would be better and, consequently, worth pursuing. Second, this innovation should contribute to economic growth and development in general, while not damaging the environment, at least not beyond human control. The main goal seems to be prosperity, though peace is a necessary condition. Order and stability are inherent in development, presupposed in the 'peaceful' West, emphasized in China (at home and in its Silk Road programme) and specifically relevant for the monsoon area, where terrorism and failed states are part of the picture.

1.2 Responsible Innovation

Innovations are 'responsible' on the condition that they are co-shaped by all actors involved on the basis of information on impacts and options evaluated in terms of ethical values, or, to give the full definition from a European report, Responsible Research and Innovation is

> a comprehensive approach of proceeding in research and innovation in ways that allow all stakeholders that are involved in the processes of research and innovation at an early stage to obtain relevant knowledge on the consequences of the outcomes of their actions and on the range of options open to them and to effectively evaluate both outcomes and options in terms of societal needs and moral values and to use these considerations as functional requirements for design and development of new research, products and services [1].

Responsible Innovation has been further detailed by distinguishing four dimensions:

- Anticipation of the consequences of a new technological project.
- Reflection on the moral, institutional and other backgrounds of a technology.
- Deliberation with the stakeholders.
- Responsiveness to the issues, concerns and values of the stakeholders [2].

Responsible Innovation includes a variety of values and, consequently, a key issue is how to deal with diverging, competing and conflicting values. Several methods are available for this, sometimes divided into optimizing methods (Efficiency or effectivity as dominant value/s, [Social] Cost–Benefit Analysis, Multi-Criteria Analysis) and sub-optimizing methods (Satisficing, Logic/Reasoning, Value Sensitive Design [looking for win–win designs] and Diversity) [3]. These methods are hard to use, especially in non-Western situations, where actors are especially or also divided and driven by cultural factors [4]. Under these circumstances, a process (or procedural justice) method seems to be better than one of the methods above, which all require information, expertise and rational considerations, except the last one: diversity. A dialogue method focused on the design and realization of a specific solution seems to be most practical.

Responsible Innovation emerged in Western Europe, but it is spreading all over the globe, including China, where there is a rising interest in it, at least from scholars [5]. The main problem here is the possibilities for public participation, but better education, the increasing availability of more information through mass and social media and the growing living standards and conditions in general are bringing change and improvement [6, 7]. Responsible Innovation, however, does not automatically imply or lead to economic growth and development. This requires more, although visions are divided on what is exactly needed and for what purposes.

1.3 Viewpoints on development

1.3.1 The concept of development

Historically speaking, development refers to the transformation from a traditional, agrarian society to a modern, industrial one, from poor to rich. This process started in Great Britain in the 17th and 18th centuries and spread from there first to Europe and the Anglo-Saxon colonies and then to other regions in the world. While industrialized countries moved to other stages of development (services, Sustainable Development, knowledge society), present-day developing countries are still in the earlier stages of industrialization, which constitutes the challenge of development.

Development includes economic growth but it is much broader than that. It comprises the whole transition from a traditional society to a welfare state, with relatively high standards of living (prosperity), usually relatively small differences in income (shared prosperity) with social benefits for the needy and with generally accessible services in domains like healthcare and education (welfare services). This transition can be called 'human development'. While economic growth is measured in terms of GDP (Gross Domestic Product) per capita, this development is measured through the Human Development Index, based on the work of Amartya Sen [8]. The latter includes, in addition to economic growth, livings standards, health, life expectancy and education, thus forming a much better policy and evaluation framework in the transition from a traditional to a welfare society than the GDP per capita.

In response to environmental and climate challenges, 'sustainable development' came to the fore, initiated by the famous Brundtland Report [9], which called for considering future generations as well the poor within our own time (intergenerational and intragenerational solidarity, respectively; this increased attention fits in with the so-called environmental Kuznets curve). Sustainable Development is known for its three Ps: People, Planet and Profit. It is a process of economic growth, considering ecological possibilities and promoting social equity, ultimately based on human rights. All of this has introduced new sets of criteria for economic growth and development, including low carbon emissions or even carbon neutrality and further reduction of income differences within and between countries. The present 17 United Nations Sustainable Development Goals, which have replaced the earlier eight Millennium Development Goals, provide generally accepted guidelines [10] (see Figure 1.1).

When it comes to development, Sustainable Development is still the most commonly used and the most applicable concept, with acceptance of the

Figure 1.1: United Nations Sustainable Development Goals. (Source: Wikimedia Commons, 2018 [49].)

moral-ethical values inherent in it. With its comprehensive character (as a 'container concept'), it covers most other terms, including inclusive development, the youngest branch on the development tree, emphasizing, in the wake of social sustainability, the inclusion of all people in the economic processes of growth and distribution [11].

This specification and broadening of 'development' are the products of a historical process, in which scholars redefined the concept based on new insight in the historical course of things and new moral standards. In this process of conceptual development, means and goals are sometimes confused when moral-ethical goals are presented as necessary conditions. The goal we have in mind in this context is shared or inclusive welfare or prosperity. The debate in development circles, however, is especially about the means, the conditions: what is needed to get there with two general models.

1.3.2 Economic growth and development: The Western model

Economic growth and socio-economic development in the West have been empirically investigated and theoretically described by Walt Rostow [12]. In his view, traditional societies are static, their economies in stagnation. Economic growth and development break this situation, which results in a society characterized by continuous economic growth. This process of change and transformation takes place through a number of stages, in which the 'take-off' stage is conspicuous. Along the way, a traditional agrarian society develops into a modern industrial society, characterized by high mass consumption and welfare services. This transition, which has a self-sustained character, is motored by private capital investments enabled by savings and profits. During this process, growth figures can be high, while the income distribution at first becomes more unequal but later less so, so that eventually the whole society shares in the rising standard of living (Kuznets curve). Rostow has also contributed to Modernization Theory, especially when it comes to original versions that claim that economic growth and development are conducive to democracy.

This model has been applied to development efforts based on the successes of the post-war Marshall aid to Western Europe. Capital investments are emphasized, and poor countries, lacking the necessary savings, can be supported with foreign aid. Jeffrey Sachs has put forward a modern version of this model and advocates massive foreign aid to developing countries [13]. His work was and is the basis of the Millennium and the 17 Sustainable Development Goals [10].

Historically, however, savings, profits and investments have not always automatically resulted in growth and development. Also, capital investments and economic growth may not be beneficial to every group in society and could even promote or contribute to 'underdevelopment,' besides development, within countries and internationally [14]. Consequently, other economists have added necessary conditions, including entrepreneurship and all kinds of institutional and cultural factors (e.g. David Landes [15]).

Likewise, Sachs' visons are dominant in the 'official' international development community, but, nevertheless, are a topic of hot debates. Critical economists proclaim the failure of post-war development aid and come up with alternatives. William Easterly disapproves of 'big plans' and pleads for small-scale, entrepreneurship based initiatives, focused on solving local problems [16]. Paul Collier emphasizes the importance of institutions [17], like Dambisa Moyo [18]. As an African economist, Moyo also underlines that 'the Chinese are our friend.'

Acemoglu and Robinson maintain that, in the long run, institutions necessary for development should be 'inclusive' instead of 'extractive,' beside a central state [19]. Inclusive economic institutions comprise market principles, while inclusive political institutions imply democracy (in the Western sense of free elections for a parliament and a government or, in a broader sense, public participation). The significance of the latter, however, is contested [8, 20, 21], not least because of Chinese economic successes and views.

1.3.3 Economic growth and development: The Chinese model

The Rostow model seems relevant for understanding the successes achieved by emerging economies, although the role of the government (an important institution!) has been relatively great there. China, for example, has had economic growth of around 10% in the 1980–2010 period, but only after the introduction and expansion of market principles and after earlier reforms in the Mao period. The time of high growth figures seems to be over, but economic growth is still high (so a successful take-off). Welfare services have remained limited, but currently, efforts are being made to distribute the benefits of economic growth more evenly across society by targeting backward areas and groups. Motors of its growth and development have been a series of 'development zones' where increasingly high-tech industrial enterprises have been set up in collaboration with foreign enterprises focused on export, following the example of (and with much capital input from) Hong Kong.

With regard to the world's environmental and climate threats, China seems fully devoted to fighting these shoulder to shoulder with most other countries. The signs were shown when China signed the 2015 Paris Agreement aimed at keeping global warming below two degrees compared with pre-industrial levels. Specifically, it is a forerunner when it comes to the production and adoption of electrical cars, the development and use of solar energy, the construction of eco-cities [22] and the electrification of ports [23, 24].

However, in terms of the importance of institutions and especially of inclusive ones, the debate among scholars on the best development strategy seems to go in a direction different from the Chinese approach. This appears from a comparison of the (1989) Washington Consensus versus the (approximately 2010) Beijing Consensus, as represented in Table 1.1.

Table 1.1: Washington Consensus and Beijing Consensus [13, 25].

Washington Consensus	Beijing Consensus
Free markets	Major role for government
Strict loan conditions	No conditions
Local enterprises and employment	Chinese labourers, enterprises
Technology transfer	Chinese technology
Democracy and good governance	State control and no recognition of Taiwan
Liberalization and deregulation	Support for China in the UN and internationally
Limited time for politics	Political stability

In fact, the Washington Consensus has been further detailed and revised by various present-day development economists and other scholars, including Jeffrey Sachs, William Easterly, Paul Collier, and Daron Acemoglu and James Robinson.

China proves that a booming economy, followed by the establishment of a modern welfare society, is possible without a national democratic system (or inclusive political institutions in general), and its foreign policies reflect this experience [26]. In doing so, China simply addresses its staying in, especially, the 19th century, restoring its earlier advanced status and position in the world [27, 28].

Whether democracy will follow upon economic growth in China, as traditional Modernization Theory seems to claim, is an open and interesting issue. In some countries, democracy backfired, for instance, in Sri Lanka. Still, the democratic countries in the monsoon area seem to be among the most developed ones, especially India and Indonesia. However, these countries are no enthusiastic partners in the Silk Road programme. India is even a rival sea power in the area, together with Indonesia more or less joining the United States in a coalition to mitigate China's increasing power in the Indian Ocean [29].

Generally, experiences in developing countries show a mixed picture in this respect, and even the history of the West shows that democracy and a growth economy in combination with a welfare society developed more or less in interaction with each other, which means that the theory must at least be refined [30]. If there is a 'first cause,' a strong (instead of a soft) and effective state seems to be a likely candidate, and not democracy or growth [31–33]. It seems safer, however, to assume, as Francis Fukuyama does, that political, economic and social developments should go together in forms and degrees suitable to specific societies and their histories. However vague this may sound, it always includes some form of stakeholder and public participation and support!

China itself also shows a mixed picture in terms of development and inclusiveness, and it seems that the domestic Chinese development strategy (China considers itself a developing country) differs somewhat from its international development strategy, although technology and economics prevail here as well [26]. In tackling its increasing environmental problems, China is pioneering Sustainable Development (see previous examples). Some degree of public participation is visible here as well, sometimes only in its most elementary form of demonstrations. It seems that the Chinese government is increasingly under pressure from public opinion to go sustainable [34].

Still, the Chinese approach, underlying the OBOR initiative, does not require inclusive development goals from its partners overseas. It puts the responsibility for inclusiveness on the collaborating states, apparently assuming – in line with versions of Modernization Theory and the theory of Kondratieff waves (see below) – that development will nevertheless result from large-scale port (and other) projects as engines of economic growth [26].

1.3.4 The role of technology

Generally, the construction of roads, railways and other infrastructural works provides a necessary condition for the development of a country, and Western economists from Walt Rostow to Jeffrey Sachs advocate such works, while historical and present practices in China and elsewhere seem to confirm their importance. Although not every social group will profit from these public works, ultimately, nobody seems to doubt the significance of large infrastructural works.

In a broader sense, technology is crucial to development. Classical economists like Thomas Malthus, David Ricardo and even Adam Smith underestimated the impact modern technology would have and, consequently, were afraid growth would be untenable in the long run; it is the reason economics was called the 'dismal science.' Joseph Schumpeter called attention to the importance of new technologies besides entrepreneurship. He showed and argued that new technologies were not only at the root of the Industrial Revolution in Great Britain (e.g. the steam engine) but were also the drivers of a number of subsequent waves of economic growth and development, sometimes named Kondratieff waves, after the economist who initially found them.

Nowadays, technology is a major part of 'innovation systems,' which describe, analyse and evaluate the spread of innovations in society and their contribution to development. They are also used to support innovation and development policies. Various types have been specified, including technological and national innovation systems. The necessary embedding of an invention in society requires attention and action in relation to the development state of a particular technology and involves an entrepreneurship based technological innovation system [35]. On a higher level, there should be a functioning regional or national innovation system that can promote socio-economic development [36].

1.3.5 Peace and prosperity

In view of the Silk Road countries, finding a power balance between different social groups, especially ethnic minorities, may be more important than democracy. Low-income groups (differences in income are usually high in poor countries) and especially ethnic minorities are often victimized in economic and political modernization programmes of states in transformation [29].

Although centralization is crucial to guarantee a strong state, democratic or not, ethnic and other diversity may require a certain degree of decentralization as well, with strong positions for minorities and other marginal groups. This poses a major challenge for the countries engaged in OBOR, as many of them are in the category of weak if not failed states, often accompanied by chaos and violence, including terrorism (for example, Pakistan, Burma/Myanmar and Bangladesh, see [37]). The top priority of development should therefore be peace and stability, along with economic growth and ecological solidity.

1.3.6 Conclusion

In this volume, large port construction or expansion projects or other projects in large ports will be considered as innovations and they will be analysed and evaluated from the perspective of Responsible Innovation, not only to assess whether they meet its criteria but also to determine which improvements might be possible.

Technology and infrastructure are important to economic growth and development and that certainly applies to large port projects. Economic growth is self-evident, but development comprises many aspects, referring to values and conditions. Alongside economic growth and development, we will speak of Sustainable Development in this volume.

Inclusiveness is part of the dimensions of Responsible Innovation (stakeholder involvement) as well as the demands of development (it is a defining characteristic of inclusive development!) and it seems to be essential not only morally but also practically. Its specification, however, is a topic of debate and finding local solutions (under the heading of 'diversity,' see [3]). In the countries and regions of the monsoon area such inclusiveness as well as the institutions to accommodate this are particularly significant when it comes to ethnic and other minorities. Ultimately, the main thing seems to be that people are somehow involved in new technological and economic developments and, in particular, that they share in the benefits: shared welfare or inclusive prosperity, seen here as the main development goal.

This volume will investigate a number of cases as to whether OBOR host countries and their populations have a chance to profit from large-scale port projects in terms of economic growth and development, or whether these new ports remain isolated zones, 'cathedrals in the desert,' as the Italians used to say about industrial complexes built in their backward South [38], or even if they contribute to marginalization and impoverishment. In addition, ecological issues receive a lot of attention, actually most attention. Here and there, stabilizing effects that might materialize in regions and countries will be discussed as well.

1.4 Diffusion, circulation and leadership

Historically, the Silk Road was not only the scene of trade between China and Europe but, more generally, it enabled the exchange of goods, people and information between East and West [28, 39]. Consequently, the historical Silk Road supports the claim of the 'diffusionists' in cultural anthropological circles that diffusion constitutes a much more powerful driver in development and growth than independent inventions and innovations, although the last position has been more attractive to modern anthropologists seeking to emphasize the uniqueness of 'their' people, village or other group of study [40]. This discussion also raged – and rages – in other disciplines, for example, within history.

In the history of technology and world history, for instance, many researchers and authors have emphasized the unique character of the West (e.g. [15]). Although with the emergence of China and a growing body of empirical data from all over the world, this perception of European or Western 'exceptionalism' is at least being mitigated, making a place for the broader view of non-Western world history. The flow back and forth of goods, beliefs, technologies and even diseases between East and West ultimately paved the way for the emergence and expansion of the West. As Francis Bacon already noted, three inventions were crucial to this: gunpowder, printing and the compass. The rest of it, the consequent processes of modernization and Westernization, is history. China, however, was not only far more developed than Europe in the past [28], but it also kept track of the West for a long time, until China's disastrous 19th century [41].

A splendid example showing this is the recent work about Koxinga, a Chinese warlord fighting the Manchus in their conquest of China in the 17th century, using gunpowder. The broader point is that the Chinese not only invented the cannon but were also quick to follow Western improvements in gunpowder technologies, neutralizing any Western advantages. Because of this and his outstanding leadership qualities and skilful war tactics, going back to the work of Sunzi ('war is the art of deceiving'), Koxinga beat the Dutch United East Indian Company (VOC) and drove them from Taiwan [42, 43].

The famous world historian William McNeill put all his money on diffusion and his son has remained loyal to this position [44, 45]. They support and elaborate Bacon's argument and show how the 'rise of the West' has been enabled and stimulated by inputs from China. In their view, world history shows a series of subsequent centres of development and innovation, with new centres building upon older ones and always absorbing their technologies and ideas, thus continuing existing pathways. They show and discuss that China was dominant between 1000 and 1500, as such preceded by classical Mediterranean civilizations, India and the Islamic world and followed by Europe and the rest of the West. For the future, the authors seriously consider that China will become the new leader of innovation and growth in the world, thus recovering its earlier central place in world affairs, as also suggested by the fourth Needham question [46]. In that respect, China is doing the right thing: it is learning from the West [20], though currently, the tables seem to be turning.

Whatever the future may be in terms of shifting centres of world history, the circulation of goods, knowledge, people and technologies will remain crucial to development and growth everywhere, and it can be expected that the 21st Century Silk Road will act as an engine in this, both as a transportation network and as a platform for the exchange of best practices and information [47]. However, through and beyond this initiative, China is aspiring to a broader leadership in the world [27], economically, politically and, in line with its philosophical and other traditions, also morally. And as Confucius and the other Chinese sages used to do, setting a good example is crucial. Whether the country also shows leadership in Responsible Innovation and Sustainable Development when it comes to ports is a question that will be addressed in this book, which first and foremost, however, is about practices and ideas regarding port innovation and development along the Silk Road, serving the exchange of information rather than something else.

1.5 Lead concepts and questions

In applying Responsible Innovation and Sustainable Development to deep-seaport construction and port-motored economic and other developments, there are two roads to wisdom:

1. The theoretical road – exploring and building visions and models, while carrying out supportive scientific research.
2. The practical road – exploring practices and drawing lessons from that.

The second road is followed by the OBOR initiative, which aims to both realize public works and exchange information, focused on improving port construction in conformity with the OBOR goals. Scholars follow the theoretical road. The ideal situation, of course, is one of interaction and mutual learning. Scientists can learn from the practitioners, and their research and analytical results can be introduced into building practice, so that in the end both paths merge, with the academic path perhaps being a detour, but a relevant one (see [48]).

The mission of this volume is to be part of this detour: its contributions are theoretical, analytical and evaluative based on desk and empirical research. Case studies of port development constitute a major source of information, but theoretical and especially methodological explorations are also part of the game.

The relevant research questions for empirical research into a number of cases of port construction and expansion projects have been:

1. What is the problem to be solved by the port project?
2. What is the proposed solution, i.e. how does the project address the problem?
3. Who are the stakeholders?
4. What are their values, besides their interests and goals?
5. What are the direct economic, social and ecological impacts?
6. What is the indirect development impact of the project?
7. How do the impacts relate to the goals of the project?
8. How do the impacts relate to stakeholder values?
9. How do the impacts relate to key sustainability values, e.g. economic profit and growth, justice or inclusive development and ecological soundness?
10. What are the dilemmas or conflicts in terms of values?
11. How are these addressed?
12. How is the project shaped and implemented?
13. Which (practical) improvements are possible and necessary?
14. What can we learn (academically) from this project?

These questions enable analysis and evaluation of the port projects involved from three perspectives relating to: project goals; stakeholder values; and socio-economic development in general. They also enable the suggestion of practical improvements of the projects concerned.

The selection of port projects has been made on the basis of interest, expertise and relevance. Some ports are discussed several times, especially the Dalian and Shanghai ports, but their importance is justified and the perspectives differ.

Not all these questions will be dealt with in every case chapter, for practical reasons of a lack of data and a particular focus, but also because all research has not been guided in a conceptual and theoretical framework. Actually, Responsible Innovation has functioned as an 'orientating concept,' each author using their own approach, although against the background of a common literature. In addition, Responsible Innovation itself, in relation to port development and Sustainable Development in general, is a subject of research and reflection, in which the case research enables the drawing of academic lessons for its content and meaning. Both concepts have been used in relation to each other, and combined and applied to ports they constitute Responsible Port Innovation. Consequently, this book aims to contribute, however modestly, to a more sustainable and responsible 21st Century Maritime Silk Road as well as a conceptual and theoretical understanding of Responsible Port Innovation.

References

[1] Jacob, K., et al., Options for strengthening responsible research and innovation: Report of the expert group on the state of the art in Europe on responsible research and innovation. Brussels: European Commission, 2013.

[2] Stilgoe, J., Owen, R., Macnaghten, P., Developing a framework for responsible innovation. *Research Policy*, 42, pp. 1568–1580, 2013.

[3] Poel, I. van de, Values in Engineering Design. In: Meijers, A., (ed.), *Philosophy of Technology and Engineering Sciences*, Elsevier, Amsterdam, pp. 973–1006, 2009.

[4] Arliana, R., Mitigating Value Conflicts in Large Infrastructure Projects in Indonesia: The case of National Capital Integrated Coastal Development (NCICD). Master's thesis, Delft University of Technology, 2018.

[5] Wang, J. & Brey, Ph., (eds), *Theory and Practice of Responsible Innovation*. Science Press Ltd., Beijing, 2019. (In Chinese.)

[6] Ravesteijn, W., He, J. & Chen, C., Responsible innovation and stakeholder management in infrastructures: The Nansha Port Railway Project. *Ocean & Coastal Management*, 100, pp. 1–9, 2014.

[7] Song, X., Ravesteijn, W., Frostell, B. & Wennersten, R., Managing water resources for sustainable development: The case of integrated river basin management in China. *Water Science and Technology*, 61, pp. 499–506, 2010.

[8] Sen, A., *Freedom as Development*. Oxford: Oxford University Press, 1999.

[9] World Commission on Environment and Development (WCED), *Our Common Future*, UN, 1987.

[10] Wikipedia, Sustainable Development Goals, https://en.wikipedia.org/wiki/Sustainable_Development_Goals, accessed on: 24 October 2018.

[11] Schulte Nordholt, N.G., In- of uitsluiting: dat is de vraag, Inaugural Address Twente University, Enschede, 2004.

[12] Rostow, W.W., *The Stages of Economic Growth: A Non-Communist Manifesto*. Cambridge: Cambridge University Press, 1960.

[13] Sachs, J., *The End of Poverty: How We Can Make It Happen in Our Lifetime*. London: Penguin UK, 2005.

[14] Frank, A.G., *The Development of Underdevelopment*. Boston, MA: New England Free Press, 1966.

[15] Landes, D. S., *The Wealth and Poverty of Nations: Why Some are So Rich and Some So Poor*. New York & London: W. W. Norton & Company, Inc., 1998.

[16] Easterly, W., *The White Man's Burden: Why the West's Efforts to Aid the Rest Have Done So Much Ill and So Little Good.* London: Penguin, 2006.

[17] Collier, P., *The Bottom Billion.* Oxford: Oxford University Press, 2007.

[18] Moyo, D., *Dead Aid: Why Aid Makes Things Worse and How There Is Another Way for Africa.* London: Penguin, 2010.

[19] Acemoglu, D., Robinson, J.A., *Why Nations Fail: The Origins of Power, Prosperity, and Poverty.* New York: Broadway Business, 2013.

[20] Mahbubani, K., *The New Asian Hemisphere: The Irresistible Shift of Global Power to the East.* New York: PublicAffairs, 2009.

[21] Kroesen, O., this volume, Chapter 2.

[22] Lu, H. et al., this volume, Chapter 4.

[23] ABB's shore-to-ship technology offers sustainable solution for Chinese port of Dalian, ABB, https://new.abb.com/news/detail/43944/abbs-shore-to-ship-technology-offers-sustainable-solution-for-chinese-port-of-dalian, accessed on 23 June 2021.

[24] Li, Z. & Ravesteijn, W., this volume, Chapter 5.

[25] Zhao, S., The China Model: Can it replace the Western model of modernization? *Journal of Contemporary China*, 19(65), pp. 419–436, 2010. DOI: 10.1080/10670561003666061.

[26] Zhang, Y., this volume, Chapter 3.

[27] Pomeranz, K., *The Great Divergence: Europe, China & the Making of the Modern World Economy.* Princeton, NJ: Princeton University Press, 2001.

[28] Needham, J., *Science and Civilization in China*, Cambridge: Cambridge University Press, 1954–2015.

[29] Kaplan, R.D, *Monsoon: The Indian Ocean and the Future of American Power.* New York: Random House Incorporated, 2011

[30] Mokyr, J., *The Enlightened Economy: An Economic History of Britain 1700–1850.* New Haven, CT: Yale University Press, 2010.

[31] Wertheim, W.F., *Third World: Whence and Whither? Protective State versus Aggressive Market.* Amsterdam: Aksant Academic Publishers, 1997.

[32] Fukuyama, F., *The Origins of Political Order: From Prehuman Times to the French Revolution.* Farrar, Straus and Giroux, 2011.

[33] Fukuyama, F., *Political Order and Political Decay: From the Industrial Revolution to the Globalization of Democracy.* London: Macmillan, 2014.

[34] Contribute to Belt and Road Initiative, Promote Connectivity, Silk Road Fund, http://www.silkroadfund.com.cn/enweb/23773/index.html, accessed on: 5 June 2018.

[35] Hekkert, M.P., Suurs, R.A., Negro, S.O., Kuhlmann, S., Smits, R.E., Functions of innovation systems: A new approach for analysing technological change. *Technological Forecasting and Social Change*, 74, pp. 413–432, 2007.

[36] Freeman, C., The 'National System of Innovation' in historical perspective. *Cambridge Journal of Economics*, 19, pp. 5–24, 1995.

[37] Michalakeas, D. & Garay Garcia, O., this volume, Chapter 6.

[38] Ravesteijn, W., Milena Revisited. MA thesis, University of Amsterdam, Amsterdam, 1979.

[39] Frankopan, P., *The Silk Roads: A New History of the World*. London: Bloomsbury Publishing, 2015.

[40] Harris, M., *The Rise of Anthropological Theory: A History of Theories of Culture*. New York: Thomas Y. Crowell Company, 1968.

[41] Pomeranz, K., *The Great Divergence: China, Europe, and the Making of the Modern World Economy*. Princeton, NJ: Princeton University Press. 2009.

[42] Andrade, T., *Lost Colony: The Untold Story of China's First Great Victory over the West*. Princeton, NJ: Princeton University Press, 2013.

[43] Andrade, T., *The Gunpowder Age: China, Military Innovation, and the Rise of the West in World History*. Princeton, NJ: Princeton University Press, 2017.

[44] McNeill, W.H., *The Rise of the West: A History of the Human Community*. Chicago, IL: The University of Chicago Press, 1991.

[45] McNeill, W.H. & McNeill, J.R., *The Human Web: A Bird's-Eye View of World History*, New York & London: W. W. Norton & Company, Inc., 2003.

[46] Ravesteijn, W., this volume, Preface.

[47] Frankopan, P., *The New Silk Roads: The Present and Future of the World*. London: Bloomsbury Publishing, 2018.

[48] Hong, P., Jointly Building the 'Belt and Road' towards the Sustainable Development Goals, 2016. DOI: 10.2139/ssrn.2812893/.

[49] Wikimedia Commons, Sustainable Development Goals, https://commons.wikimedia.org/wiki/File:Sustainable_Development_Goals.jpg, accessed on: 6 December 2018.

CHAPTER 2

East is East and West is West . . . and the two finally meet in the Indian Ocean: Competing powers in the Indian Ocean, and the power of self-restraint

Otto Kroesen

Delft Centre for Entrepreneurship, Delft University of Technology, The Netherlands

Every step in the direction of organizing the world's economy will have to be bought off by a great number of tribal reactions.
E. Rosenstock-Huessy
Out of Revolution: Autobiography of Western Man (1938) [1]*

Abstract

The Indian Ocean is the scene of increasing competition between different nations, power blocs and civilizations. The ongoing power game may become more serious than the word 'game' suggests. Can the different stakeholders and competitors only deal with each other from the outside as blind forces in space, or could they find ways to balance the competition by forms of cooperation and recognition? This chapter attempts to find the resources to do so in the moral and spiritual traditions of the different stakeholders and competitors involved. Within the different traditions inhabiting the world, there have always been sets of moral values that cultivated self-denial and self-restraint. These traditions from the East and the West will be explored. They may support the Responsible Innovation approach to be used within the framework of the One Belt, One Road initiative in going beyond the level of rational self-interested competitors, trying to optimize their outcomes at the expense of each other. The value traditions of self-restraint make the stakeholders appear in another light by giving them history and a human face. It is the respect for what is most valuable – and also vulnerable – in each other's traditions that creates room for mutual recognition and more responsible and inclusive solutions.

*Rosenstock-Huessy makes this statement in view of the Russian Revolution, the World Wars and Nazism. He expects that what he considers as the fundamental problem behind these conflicts – how local traditions/identities are respected and yet integrated into a global economy – will haunt us for generations to come ([1] p. 718).

Keywords: Value-sets, self-restraint, moral-spiritual traditions, tribe, empire, civil society, state, Responsible Innovation

2.1 Introduction

More than 20 years ago, Samuel Huntington in his book *The Clash of Civilizations* [2] predicted that in the future the dividing lines would not be between the free world of capitalism and the closed world of communism, but between the different civilizations of East and West. Since then, the economic and military powers of China and India have been on the rise, while the economic and military power of the United States seems to be comparatively diminishing. The One Belt, One Road policy of China exemplifies the increasing power of China in the region. Economic power translates into political power and subsequently into military power. Competition is there. Conflict may be lurking around the corner [3]. But conflict between what and who? Even economic power is not merely a matter of self-interest. Such self-interest happens to be already morally framed in some particular way. The institutions and the actions of these powers express sets of values and ways of life. Each of these state institutions considers these as viable and valuable export products. They may function to the benefit of all, but if necessary, they are also worthwhile to defend, even with military means. The civilizations and nation-states inhabiting the global world do not want naked domination, but all of them represent a particular and specific value-set, if not meant to dominate, then at least to be respected and appreciated.

Lack of respect may be more dangerous than mere lust for power. Even the Buddhist monks in Myanmar perceive their reflective meditative traditions as endangered by upcoming Islam. The Rohingya have to suffer from it. Border conflicts between India and China become more severe if both of them fear a loss of face for nations they control (Tibet) or are aligned with (Nepal). The tribes in Afghanistan and Pakistan consider their traditional lifestyles and local authorities (the rule of big men) as being threatened by the intrusion of Western values and production methods and they cling to fundamentalist Islam, turning it into a means of violent aggression. The United States and its Western allies are anxious about the Chinese exploitation of African resources and inclined to perceive the One Belt, One Road policy as just another means to promote an extractive top-down regime. But the Chinese, in turn, may just as well cling to such strict control from the top out of fear of supposedly destructive Western values, which are perceived as the cause of loss of cohesion within its society and of promoting centrifugal forces in and around China. They may at the same time consider the slow expansion of military influence (islands created in the Pacific) and development of harbours in the Indian ocean as natural and legitimate (see Figure 2.1). There is a long tradition of China making its satellite states comply with its policies by leaving them no other

Figure 2.1: East meets West in Gwadar, Pakistan. The peninsula is the location of Gwadar Port. (Source: NASA, 2018 [50, 51].)

option – merely by peaceful means. How could cooperation be established otherwise – from the Chinese point of view – but from the gravitational force of its Imperial Centre? That's how it always worked. However, for the United States and its Western allies, the right of self-determination and freedom of public opinion are just as natural a means of creating a support base for legitimate policies – whereas in the eyes of other powers this may already be perceived as Western domination, endangering their traditional values. The self-evident behaviour of one party is perceived as a threat and a danger by the other. Even our 'natural' inclinations are value-laden. How do you find agreement if the values we adhere to make us differ on the legitimate political and value-laden institutions and instruments for reaching such agreement?

In this contribution, I intend to focus on these values and on the institutions in which they are embedded. The mere fact that conflicts are value-laden doesn't make them less violent. What values are at stake and can these values open up towards each other and speak to each other – do they have something to say to each other? This question requires a grand historical telescope, bringing historical events from quite different origins into one perspective. This is necessary to get to the inner core of these traditions, or soul, if the reader accepts the word. The word 'soul' expresses the moral core of a tradition of which the living generations somehow cannot let go, even to their own surprise [4]. The question of whether these value-sets can open up to each other leads to the further question, whether these different institutions and value-sets can find a source for self-restraint in their own and also in each other's traditions? To an extent, an exchange of perspectives already took place, since already in large-scale Western institutions of state, civil society and market formation have been copied in some or other version by all the nation-states in the East [5]. One may even wonder whether the arrow may also point back: can Western civil society and institutions learn something from the value-sets from the East? Is such a thing also taking place? Can the ongoing power game be turned into cooperation or can the competition at least be balanced by forms of cooperation? Can the competition have its way by peaceful means?

In Section 2.2, I will draw attention to the concept of value-sets. Civilizations are not led by only one or even by some different values but by sets of different and often even contradictory values, which pertain to different spheres of life. In their private life, people may maintain different values than in public. Different ages, even different sexes may be attached to different values. In Section 2.3, I will give an overview of the value-sets originating in the East and in the West and how these values shaped their histories and institutions. Section 2.4 is about the interaction between these traditions from East and West and how values and institutions from the West have merged with the Eastern inheritance, particularly in India and China. Section 2.5 focuses on different ways of dealing with Responsible Innovation resulting from these histories. There I will also consider the impact of technology on societies both in the East and West. The conclusion, Section 2.6, answers the question of how the different values and institutions from East and West can be oriented to the common goal of living together on a small planet. Can these different value-sets reinforce each other and be realized simultaneously, or are they exclusive and can they only be realized at the cost of one another?

2.2 The concept of value-sets and spheres of ethics

Many studies have been conducted into the history of values [6–8], but little research has been done on the way we combine different values and how we move from one value to the other, how we polarize opposing values and establish cooperation between contradictory values. To give an example of the latter – different values are often distributed between males and females. The male may be occupied with effective action and oriented to the outer world, while the female to safety and care in the inner world. The more they recognize each other's contribution in raising their children and follow each other in alternating between care and efficient action, the more they can deploy an interplay of values that offers an enabling environment for their children. They do not only interrupt each other but contribute to one goal by mutual correction and interference if only they find the right rhythm [9]. Their values can interpenetrate. For instance, if the father chooses a car because it is cost-effective, the mother because it is safe, the son because of its speed, and the daughter because of its beauty, this family may have chosen just the right car, realizing those four different values in one stroke. It is important to know when to intervene and when to stop. The father should know when to stop his being efficient and allowing, and even incorporating, values of tenderness and care. Father and mother may polarize each other and call forth each other's actions and even confirm and empower each other in their different contributions by finding the right rhythm or algorithm.

The same counts in politics. To govern implies authority and command. But it should also integrate accountability and care into its value-set. Just like there are spheres of justice in which different values count [10], there are also spheres of ethics. In many societies, people have to be careful not to criticize those higher up in the hierarchy, while people higher up in the hierarchy are allowed to criticize subordinates. Or, if one criticizes those in authority, it should rather be done by giving a hint or by expressing it in a sub-sentence or between the lines.

If we analyse, for instance, the algorithm of individualism more closely, it appears to have some neglected dimensions. Allegedly, Western societies are individualistic. And indeed, in the public sphere, every individual is allowed to express an opinion. At face value, everybody is doing as he or she pleases. And this impression is reinforced by the official discourse on individual choice and self-realization. At the same time, however, in the economic sphere, there is a solidarity system, the social security system, which is an anonymous system guaranteed by state bureaucracy and the tax system. In many non-Western countries, it is the other way around: it is more difficult to freely give your opinion in politics, but in the economic sphere people are thrown back on themselves. Both societies know of the values of solidarity and individualism but in a different order or sphere. In order to understand societies and politics, we need this sort of distinction and therefore the concept of value-sets, and even value-orders. By value-orders I mean how values may be integrated into each other and alternate and follow each other.

2.2.1 The institutions of antiquity: Tribe and empire

The concept of value-sets and value-orders becomes more complex if the historical development of values and institutions is considered. Institutions and values are two sides of the same coin. In history, we can witness the sedimentation of institutions as geological layers of human experience and order. An old order and value may still function in the background, even if other and new institutions have become more dominant. These institutions embody different values and by interfering with each other these values are integrated into each other and alternate.

There is a layer of two sorts of institutions that is common to all cultures. One is the family. In whatever form (extended family, clan etc.), family life already entails a number of values: authority, obedience, loyalty, self-restraint, chastity, etc. The other layer of institutions consists of a hierarchical political order: chiefs, kings, emperors. We have to differentiate between tribes, which live from the authority of the ancestors, and empires, which are ruled by the necessities of the cosmos, exemplified by the stars [11, 12]. The Pharaoh was in command as Son of the Gods on top of a huge hierarchy. His rule was identified with the rule of the stars moving slowly and determining the seasonal rhythm of agricultural societies. This later development of empires in turn gave a special twist to the old tribes. They changed under the influence of empires. A king, for instance, is still a father figure, calling for obedience, but now with much more power, although at a distance and not as personalized as before, more functional and more hierarchical. Hierarchy and scale result in professional specialization, labour division and more emphasis on labour and time management. Hierarchy and labour division, by being highly productive, also introduce consumer values and welfare. That was already the case with Egypt along the shores of the Nile [12]. But the king in a smaller empire (the word is derived from kin) could more easily also embody the traditional authority of the father figure of the tribe. All societies have complicated silent knowledge of the algorithms of such alternating values, imbricated in each other, impinging on each other.

These value dimensions are closely related to the system of political rule and religion. No tribe would differentiate between a secular moral system (respect for the elders) and its religious system (respect for the ancestors). That values as secular can be separated from religion is quite a recent development. It has only been 700 years in the West and it is still uncertain whether it will hold, especially because in the end values like justice, loyalty and responsibility always turn out to be somehow sacrosanct themselves [13]. One can doubt whether people will be ready to sacrifice their lives just to maintain a secular set of values. If sacrifice is required, religious language may become unavoidable [14].

Even in the era of technology, nation-states and power blocs have inherited the value systems and institutions of tribe and empire, albeit on a more sophisticated level. That means the core values of tribe and empire are still quite alive in these modern formations. They tend to limit the moral community to their members only and act as blind forces towards the world outside. They tend to be exclusive (like the tribes of old) and exploitative (like the empires of old). This raises the question of whether modern nation-states and power blocs can cooperate in any other way than Kaplan [3] seems to be suggesting – calculating their assets and reducing the different players to forces in space to be measured and weighed against each other. Can these powers, on the contrary, find moral sources within their own and each other's traditions that enlighten them and open their eyes to feel a responsibility towards each other, or at least enlighten them to a higher level of rational self-interest? The dividing line between reflective responsibility to one's neighbour and well-understood rational self-interest may be difficult to draw after all. But without one moment of altruism and putting ourselves in the position of the other, we might not even discover our own well-understood rational self-interests [15].

2.3 Value-sets and institutions from East and West

In order to answer these questions, we have to expand our analysis even beyond the value-sets of tribe and empire as described previously. Imperial values like authority and hierarchy have been challenged by at least four traditions of criticism and withdrawal from and denial of the powers that be. They can be called countervailing values, criticizing and down-tuning those in power and giving voice to the marginalized and oppressed. On the other hand, these more critical values, if integrated into the system of rule, can also help to make imperial rule more comprehensive and inclusive.

2.3.1 Four cardinal values of self-restraint

Two of those value traditions originated in the East: Buddhism and Taoism. Two of them originated in the West: Judaism and Christianity. They are new compared with the earlier set of tribal and hierarchical values. They are even in tension with those earlier sets of values.

In summary:

1. Buddhism introduced the denial of attachment to the outward reality of struggle and competition: the eye objectifies and looks through everything. It reduces the hot steam of anger and power to nothingness. Just Maya, appearance. It does not make an impression any more [16].
2. Taoism introduced another value, to let go of recognition within society. Lao-tzu disappeared behind a mountain. To make oneself unimportant within society, not to compete for respect and status, doing nothing and withdrawing, paradoxically is a necessary condition to find the one right moment to intervene. Even such an intervention is still a form of doing nothing, of Wu Wei [17]. A movement of one's little finger may change the universe if it is done at the right moment.

3. Judaism introduced the denial of the powers that are in function of the one power of justice that is beyond them all, emphasizing a common origin of humanity, beyond our grasp. The Jews did not justify their existence by putting themselves centre stage in their myth of origin; instead, they are just one people among the many and all those peoples were created long before the Jews entered the scene (Genesis 10). This gives to all the peoples a common destiny beyond the existence of individual people. This common destiny challenges each particular myth tribes and nations would like to sacrifice to.
4. Christianity operationalized this value by giving up the claim of a total change, as voiced by the prophetic criticism of the Old Testament prophets. Thereby a path was opened of constant and gradual change in limited steps, but always in view of that ultimate reality, and always opening up a new road into the future. This is the driving force behind the revolutionary upheavals in Western civilization [18].

It may be a point of discussion how Islam, as well as a number of other traditions are included in this somewhat ideal-typical description of four 'nos.' In general, and again ideal-typically, Islam can be considered as a somewhat attenuated version of the Jewish and Christian abnegation of the powers that be. Islam looks upon history as a journey to a common destination of all humanity like the Jewish and Christian traditions and its other forerunner, Zoroastrianism. If we take this cruciform diagram of time and space in an ideal-typical sense, societies, religions, even denominations within religions, and concomitant values can be put somewhere on a continuum on one of the arrows (see Figure 2.2). They also reinforce each other and evoke each other. They need to be supplemented and corrected by each other. For instance, a critical stance towards the powers that be can become one-sided and forgo options for gradual change. In a similar way, not making oneself too important in combination with a detached attitude in the struggle for existence, although different values and qualities, may help and support each other. In China, Buddhism and Taoism were often in such symbiosis. Being without recognition, unimportant and at the same time critical may also sometimes be necessary. That has often been the attitude of the Taoists in China during peasant revolts. In a similar way, the Buddhist attitude of self-renunciation and a compassionate Christian attitude may sometimes supplement each other to prevent Buddhism from becoming too passive and the Christian striving for change too active, etc. These four spiritual attitudes and traditions should, however, better not be understood as an expression of some common denominator like compassion, although for certain in all of them some form of compassion is present [19]. It is their differences that make them meaningful for each other if they face each other, correct each other and supplement each other as real people do.

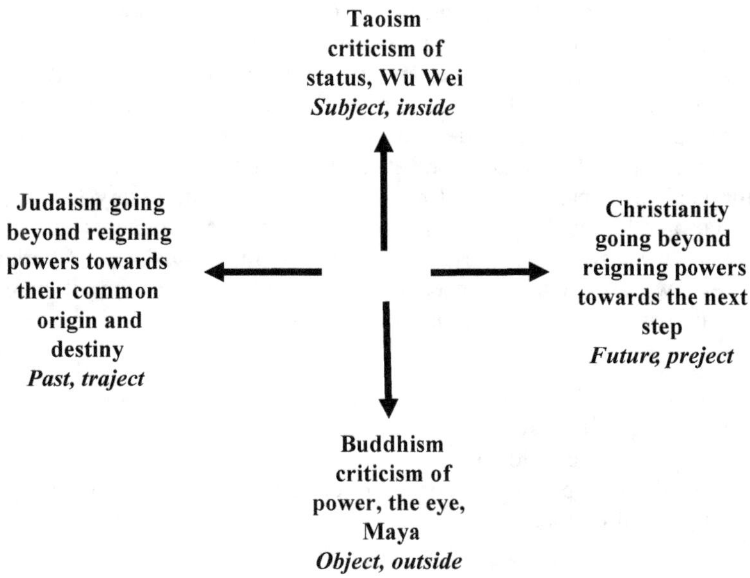

Figure 2.2: The correlation of four 'no' traditions in the East and West.

Figure 2.2 summarizes these two value-sets on the axes of space (vertical: inside society and outside reality) and time (horizontal: the tension and interaction between past and future). To the familiar space-oriented terms subject and object for inside and outside, the terms of traject and preject are added as equivalents for the time axis.

Wherever East and West meet, these two value-sets also meet. These values do not only exist in the abstract. The different civilizations and societies that populate our planet are their embodiments. Whether one calls it our unconsciousness or our soul, in some way we carry a particular formative heritage along with us. That means we are formed in a moral sense, in our feeling and outlook, by the value-sets of our cultures and religions. We are preformed by our education, our parents, our schools and authorities to react in a particular way. We cling to certain values as a solution even before reflection. We see a red light when we see particular behaviours that we cannot allow. We consider particular forms and expressions of self-interest legitimate, but others we cannot tolerate. Western nations, for instance, see a bottom-up approach of pluralistic public opinion and criticism as a natural part of making politics, but to Eastern powers, it may be perceived as threatening the coherence of society with chaos. If everybody has their own opinion, in their eyes, rule and direction become impossible.

Kings, emperors and chiefs originally may have seized power in a naïve effort of self-affirmation of a special group in the name of its group identity or gods (which is more or less the same if we understand religion in its social meaning [20]). But for sustained larger-scale rule, some form of self-restraint of the rulers was a necessary requirement. Rulers that know of no self-restraint cannot cope with political and moral systems that are different from their own and the tactical management and negotiations that are necessary to conquer and/or co-opt the stubborn self-affirmation of diverse opponents. To make large-scale rule sustainable takes the ability to give in and deal with 'otherness' in whatever way. As a consequence, in the more sophisticated forms of imperial rule, we find at least three different layers of values:

1. The values of family and clan (belongingness, loyalty, tradition)
2. The values of hierarchical rule and control (respect for authority, exercise of functional behaviour)
3. The values of restraint and denial of control as a moral and institutional countervailing force.

The axial age is generally considered a decisive turning point in the history of humanity in this respect. In the axial age, roughly from 600/500 before the Christian era, these four traditions of self-restraint mentioned emerged [21, 22]. Karl Jaspers who invented the term uses a timeframe of 800–200 BCE. All of them criticized unjust rule, but at the same time, they are of fundamental significance for larger-scale rule. Rulers could perform better if the criticism were internalized into the system of rule. It gave them a broader repertoire of action, thought and judgment. It made them more capable of dealing with difficult and stubborn people without entering into conflict and without conflicts becoming too serious. It equipped them with the ability to deal with quite different value-sets by adding four different elements of flexibility to the repertoire. Once again, to the reader who misses the Islamic tradition in this list, it should be pointed out that in Islam the traditions of Judaism and Christianity found a compromise with the traditions of tribe and empire. Islam originated as a unifying force among the Arab tribes and turned this tribal existence in the course of its history into imperial rule in the great Asian empires [23]. Mohammed considered himself to be the final reformer of the Jewish and Christian traditions. But he also left the original layers of tribe and empire more intact. His 'no' went less far. That was intentionally so. And, of course, at all times it should be part of the discussion of how far a 'no' against the powers should go at any specific time, particularly if the concrete form of such a 'no' may also consist in rebellion, revolution and civil war. Oftentimes, also Christian, even Jewish, denominations and kingdoms have followed this course themselves, for example, the Khazars at the border of Russia [23] or the Ethiopian Empire.

2.3.2 Integration of countervailing forces and voices in traditional India and China

Countervailing voices and forces were never absent in the great civilizations of India and China. I could also mention derivative civilizations like those of Japan or Java or Thailand and so on, but for this chapter, the focus is on the two big players where the traditions of Taoism and Buddhism originated. The rulers could not do without any legitimation and support of family heads, clans and tribal leaders, castes and civil servants and not to forget, the military. A wise emperor would know self-restraint and take their interests into account, gaining their support. A bad emperor would push his agenda upon them, and if such politically non-inclusive behaviour would become economically extractive [24], sooner or later the country would be in turmoil and some military leader from the inside, or rival powers from the outside, would revolutionize the country and finally conquer state power.

Mostly, it was not either/or, but somewhere in between. An emperor mainly had to buy off the opposition from the elite that he needed for support, and he was forced to do so by gaining extra revenues from the average person at the bottom of society, mostly peasants, and squeeze them or let them be squeezed accordingly. Finally, the situation would become untenable and a change of government unavoidable, which mostly would amount to a change of dynasty. A new dynasty and a ruler with a new sense of justice and inspiration for change would stand up and be more inclusive and less extractive. Typically, everybody from high to low would draw inspiration from the crisis they went through and not want these atrocities to happen ever again. In China, it seems to be the Zhou dynasty (1100 BCE) that introduced the appeal to heaven, 'Tian,' as a higher power of justice [25], over and above unjust powers – which is quite in line with the inspiration of Judaism in claiming a God beyond the existing gods: heaven should rule in order to create justice and order after a time of heavy conflicts on earth. In this way, a new beginning is announced. (Note: Buber, the famous Jewish philosopher from the first half of the 20th century has written an essay on Taoism in which the Taoist philosophers were depicted almost as Jewish rabbis, going against and marginalized by the powers that reigned, but thereby restoring the unity of action and sense of purpose in history [26].) Over time, however, when the initial inspiration fades away and people become more concerned about their daily affairs, patronage and clientelism would set in again, making the system of rule more costly and ending in another cycle of upheaval. At all times, some form of consent from the lower echelons was indispensable, and if the average person was too impoverished, he/she would no longer consider the emperor the true representative of the cosmological order of the sun and the stars. In the course of time, it might seal his fate.

2.3.3 Integrating a countervailing force and moral attitude into the system of rule

In China, a centralized system was realized by ideological support from Confucianism and criticized but also supported by Taoism. The early Han emperors were inspired by Taoism [25] and this probably made it possible for them to establish this system of centralized rule in the first place. Nevertheless, civil servants – who were inspired by Taoism, and even if they were inspired by the less controversial teachings of Confucianism – always had to watch their words. Even some criticism between the lines could lead to exile [27]. After the Ming dynasty, Confucianism largely abstained from politics and the strong imperial hierarchy by the Mogul conquerors led to petrification, turning Confucian education into little more than an instrument to reproduce state orthodoxy [28]. Where central authority held everything and everybody in its force field, the relativizing power, especially of Taoism (in a miraculous integration with Buddhism coming in from India), nevertheless remained an important countervailing force.

In India, where the family line in the castes and the Brahmin exercised strong control, it always was more difficult to centralize political power. Each attempt towards centralization sooner or later disintegrated by all the centrifugal forces of castes and small kingdoms. Nevertheless, Ashoka (300 BCE), after unifying India by military force, later converted to Buddhism [29]. One reason that Buddhism might have been attractive was the need to downplay Brahmin aristocracy, the all-important family heads, who as local rulers were a risk for his unification strategy. A religious conviction that denied the importance of castes and was critical to the costly sacrifices from which the Brahmin benefited may have suited him well.

Within the centralized empire of China, the need for recognition and the ambition to play an important role in social stratification was the primary issue of concern for the elite. In India, the struggle and competition between the different castes, often serving as a power base for larger-scale political control, was the primary issue [30]. Where Taoism by the moral attitude of Wu Wei reduced the necessity of recognition and status in the centralized Chinese system, Buddhism, by looking through the will to be and reducing the will to power to Maya, tackled the inclination towards violence and struggle. Hinduism can be considered as a form of moderated Buddhism [16]. It maintains the reduction of every manifestation of force to Maya, but it nevertheless empowers rulers and heroes to stick to their social roles even if doing so with an inner irony, a 'holy indifference' ([31], p. 20). This perhaps explains the flexibility of Indian politicians, who can powerfully show their muscles in order to get things done and, nevertheless, suddenly change their roles and course of action if so required by the necessities of the situation (see, for instance, Indira Gandhi's role in India when it turned from the Soviet Union to the United States [32]). The sudden breakthrough of 'the eye' that sees through Maya in India [18] is the equivalent of the eternal rest of Wu Wei in China that, like heaven, never changes.

It should be kept in mind that Confucianism and Taoism emerged long before, during the period of the warring states, 500 BCE. Amidst this turmoil between competing dukes and warlords, their function was often to admonish the nobility, princes and kingdoms to exercise self-restraint and make peace [33]. Their function was comparable to that of the Franciscans and other mendicant orders in the petty wars and competition among the city-states of northern Italy and elsewhere in Europe. It is probable that rule on a larger scale cannot do without such spiritual resources. I already pointed out that the Han emperors were inspired by Taoism. They unified the country and put an end to the warring states. Thereby, however, they also put an end to the freedom of movement of Confucianists and Taoists. They could no longer take refuge with the more tolerant dukes and princes if they were persecuted. From that time onwards, the freedom to speak out by wise advisors came to be seriously hampered on numerous occasions. Civil servants that were inspired by these traditions always had to be on guard.

In India, the situation was less centralized. There was no centralized exam system for civil servants, for instance. Instead, education was in the hands of the Brahmin, the highest caste. Due to their religious role, centralization by military force was more difficult because the military formed a caste of its own under the authority of the Brahmin, and therefore, often remained localized [30]. India always fell apart in different states in competition with each other. In ways comparable to China, higher state authorities always had to buy the loyalty of different subordinate groups, castes and families, and since this came with a cost, the cost was more often than not devolved to the lower echelons and castes of society. Hinduism and Buddhism, however, taught some self-restraint comparable with Confucianism and Taoism and sometimes this could improve the lot of the farmers, or it at least secured the internal cohesion of the kingdom.

2.3.4 A closer look at the ethical-religious sources of inspiration

China had peace within and enemies outside. India was always in turmoil due to the competition of the different castes and could therefore not keep its enemies sufficiently outside. This external situation corresponds to the ethical-religious inspiration of India and China respectively, as I already have put forwards. The Wu Wei attitude of Taoism is a way to make oneself unimportant in a social environment that competes for recognition and status. Everybody claims recognition from the central authorities. This need for recognition leads to stress and strain, if not the exclusion of each other. What to do in order to surmount this rivalry? Don't act! Do nothing! (= Wu Wei). Lao-tzu, the founder of Taoism, who was probably employed as a civil servant, went away and disappeared behind a mountain as the legend goes. 'The best government is the government which you don't notice that it exists' is one of his insights [17]. One might also say: don't act when it is not really necessary. If it is necessary the act will come as of itself, by non-action as it were. This ethical attitude is especially geared to enhance and maintain peace within. In a situation where everybody wants to be important, peace can only be made by the unimportant. India, contrary to China, primarily drew from a Buddhist and later Hindu ethical-religious inspiration. Just like Confucianism can be considered a moderate form of Taoism, Hinduism is a somewhat moderate form of Buddhism [16]. Always when the Buddhist looks through the busy-ness of competing and struggling warlords, states, castes, tribes and kingdoms, it sees empty-ness. In a Hindu story, a huge army is shown to the newly enthroned king. The army consists of almost endless rows of ants in marching order. It is a vast army, passing by in lengthy rows one after the other. The Godhead who shows it to the king explains: these were the kings in power before you! With the same splendour, heated self-affirmation, energy in constructing buildings, making war. With all his glamour he is just the next in the row. To look through the appearance of vitality and power of outside forces (Maya), and even to look through one's own anger, passion and virility, created room to surmount one's appetite for struggle, even if Hinduism didn't go as far in nonviolence as Buddhism. This ethical attitude is especially geared to enhance and maintain peace between groups looking at each other from the outside, as enemies in war, competition and especially in the continuous struggle between the castes themselves.

2.4 Self-restraint in the West: A universalist state and a civil society

Now what about the ethical-religious inspiration of the West? One might argue that the Western traditions of Judaism and Christianity fulfilled a similar role, but this time not in terms of space (China inside, India outside), but in time. A group of slaves, running away from the Egyptian Empire, founded Judaism. They drew their inspiration from the promise of an order of justice that was yet to be established, but not present. They drew their inspiration from an invisible religious power, beyond the oppressive visible powers, beyond the cosmo-political realm of kings and emperors that legitimate themselves as representatives of the stars and Sons of the Gods [12, 34]. Where Christianity claims Christ as the Son of God, it is a Son of God in reverse, in opposition. He is not at the top among the stars, but at the bottom, at the cross. But thereby he is robbing the Imperial Sons of the Stars of their justification and entitlement. In many respects, Christianity can be considered as a moderation of Judaism. The prophets of the Jews always position themselves in the margin of society and they do not accept compromises. The Jews did away with tribal heroes and imperial rulers and for 200 years after leaving Egypt in 1300 BCE they constituted an anarchic rule by common people and charismatic leaders. Christianity operationalized this absolute criticism by choosing a road of gradual but relentless change, always revolutionizing the existing order and creating new institutions to make progress [1]. Judaism didn't accept the cosmo-political societal framework of the empires. It also didn't accept the Christian compromises with the ruling powers, and in this sense, it remains the louse in the fur of Christian self-satisfaction throughout Western history. On the other hand, Christianity translated this radical criticism into a process of change by which step-by-step the inherited institutions of tribe and empire were changed into an open civil society (families remaining as the only remnants of the tribe) and a universalist and egalitarian state (the state, rule of law, accountability, etc., remaining as the only remnant of the holy hierarchy of the empire). Always some group took the lead and with great sacrifices instituted new forms of governance.

A most important and maybe decisive step in this ethical-religious inspiration in the West has been the struggle about realism and nominalism in the Middle Ages [35]. It was a struggle within scholastic philosophy. The realists maintained that the world was out of necessity what it was because it was created by God. Don't change it! The nominalists held that God created the world by an act of will because he is the power above and beyond, not subject to necessity. He could have created the world otherwise if he wanted so. As a consequence, the world, being considered contingent, became open to change. Of course, the nominalists in the Middle Ages were the ones propagating social change. They were on the side of the city movement, which constitutes a major step towards the establishment of a civil society [1, 36, 37]. Ordinary civilians, runaway serfs and poor people, especially during the 13th century, established cities on their own initiative from the bottom up, by just building fences, adopting a set of laws, pleading an oath on those laws and starting a community of their own. City air makes free! This movement loosened the ties of family and clan but also broke with the obedience to an all-powerful state, the nobility and so on.

The traditions of Judaism and Christianity, especially Christianity, did also provide an ethical-religious inspiration for the self-restraint of rulers. But, as with the Eastern traditions, the traditions from which one could draw inspiration were the same traditions from which criticism was also drawn. And in the West, this tradition of criticism became more institutionalized, due to the introduction of the time perspective of gradual change – although one should not forget that what we now call 'gradual' didn't go without a revolutionary effort and a lot of bloodshed.

This process of change introduced something new into history, because from now on one should not merely restore the one-time Golden Age of the past, but one could also build institutions that didn't exist at any time before. The first step was that the Pope, taking the lead of the Western church, took the nomination of the bishops out of the hands of the Emperor, and became sort of a spiritual rival Emperor. The bishops and their clerical staff were the only group of people ('clerks') that could read and write. As a result of the struggle between Pope and Emperor, a new body of law regulating their controversies was created, canon law. It codified their respective fields of authority and was considered to be binding for themselves [36]. This too is new in history. In the name of a spiritual power beyond and above all temporal powers, the latter becomes merely secular. As a consequence of the respect for a higher power beyond, Pope and Emperor were not considered to be above the law any longer, but instead, they should also be subject to the laws they instituted. This spiritual motivation created rule of law and by law [1, 36]. I already mentioned the city movement as an important step in the liberation from family and clan. It created a civil attitude of trust and cooperation between different families – 'civility.' As such, it tuned down family interests, jealousy, distrust and created a sphere of anonymous trust. The German Reformation, in turn, took the responsible individual out of the custody of the church, but it also took canon law out of the church and put it in the hands of the secular powers, the German princes – and after a while, this was taken over by the French and the British kings. English parliamentarism, emerging at the end of the 17th century, turned royal power into merely executive power by binding it to the law-giving body of Parliament. The French Revolution took this beyond the aristocracy to ordinary civilians. The Russian Revolution transferred state power to the hands of farmers and labourers, albeit as functional units of production under the guidance of a Communist Party. The Russian Revolution also exported this Western heritage to China, while the colonial power of Britain did the same with its liberal heritage to India. Two world wars created a worldwide sphere of economic globalization and competition and created a host of different systems of state and society, with different degrees of universalism at the state level and different degrees of openness and trust at the civil society level. They unified the world economically, and in doing so, they are the culmination point of the enlargement of scale that was put in motion by the cascade of European/Western revolutions [1]. The two world wars finalized this process by enforcing the dilemma on all the great powers that they have to talk to each other lest they destroy each other. The world has become small. They cannot evade each other. And due to the power of military destruction, they cannot conquer each other any more either.

2.4.1 A breakthrough in the bureaucratic system in the 1990s in India and China

Both in India and China, a tremendous change has taken place. A more universalist state has been constituted and a better-informed citizenry has been included in the political system. The economic crisis of the 1990s has sped up this process significantly. At that time both China and India couldn't do otherwise than allow some form of economic democracy or civil society. That is, they had to liberalize the economy and allow at least a degree of free competition [38]. Many forms of productive entrepreneurship were allowed to save the state from bankruptcy. The state was forced to give in and follow society. Actually, both countries thereby freed their societies from the chains of state bureaucracy. At least to an extent. If there is free competition in the sphere of the economy, society has become more open, even if there is not a completely free press or freedom of association like in China, and even if state bureaucracies still have a strong grip on the social and economic spheres like in India. Economic freedom is already a meaningful step in the direction of an open civil society. (Note: some sociologists distinguish between state, civil society and the economic sphere so that companies do not belong to civil society. Historically, it is more accurate to distinguish between state, civil society and ties of family and clan, since civil society emerged as a middle sphere of the free association of individuals between state authorities and family loyalties, see [39]. The freedom of association also pertains to the economic sphere if it isn't suffocated by the state bureaucracy.)

Landes [40] points out that the Chinese have always been more successful in trade and entrepreneurship outside China, where they were not under the control of the Emperor. There their thrift and labour made them prosper. The legalist tradition in the Chinese Empire has always been in support of strict control by the central state. In this tradition, the command of the Emperor was the only authority, and as such, legitimate without any further justification. Often, the word state is used for these strong political-religious hierarchies of imperial control without any further reflection. The Western word 'state' obscures the character of the state in the empires of the East. State, as already pointed out, means power without (religious) legitimation. Only in the Western world, where the church from the Middle Ages onwards took the lead in the spiritual life, the word 'state' could emerge as a name for the secular powers that remained after this desacralization of the imperial hierarchy [1]. It is important to note that this desacralization originated from the church. As not endowed with the Spirit, and without religious legitimation, the temporal powers (secular = temporal) are not sacrosanct any more. State means empire without religion, without, that is, the emperor or king being considered as Son of the Gods and equal to the stars. In China, religious power, the moral legitimation of rule, and state power were one and the same thing. The Emperor was the Son of the Gods, Son of the Stars, and his rule was of equal necessity as the movements of the stars. That means total control! Xi seems to be moving in that direction.

One could even go as far as to challenge the Chinese legalist tradition of state control by doubting whether it was really and wholesale supported by Confucianism. The unification of China by the Han emperors sharply limited the extent of debate within the intellectual world [41]. On the contrary, the mild irony of Confucianism and even more Taoism often expressed more criticism, even though between the lines, than was acceptable to the imperial court. Confucianism remained part of the educational curriculum for civil servants, but often not as a vivid tradition, but as a dead letter. The exams were often no more than the reproduction of knowledge [42]. The hierarchy of the state in the Ming dynasty didn't tolerate independent critical thinking and expression.

2.5 Inclusive or exclusive value-sets?

Are West and East growing together? Is it possible that their respective value-sets interpenetrate each other so that they enlarge each other's repertoire? Or should one value-set dominate the others? How can their respective value-sets and institutions withhold them from conflict and enhance their cooperation? Or can they remain critical towards each other and still work towards one and the same goal? Do they have different means at their disposal to reach the same goal and correct each other, maybe even by contradicting each other? I will here introduce the concept of Responsible Innovation used so often in this volume and discuss to what extent the different value traditions and institutions can deal with each other and find common ground within this framework in the case of the One Belt, One Road project in the Indian Ocean.

2.5.1 Responsible Innovation and inclusive institutions

Responsible Innovation is typically described as a process of four steps: anticipation, reflexivity, deliberation and responsiveness [43]. As such, it represents a process through time. New technology is anticipated and expected. It offers new opportunities in response to new imperatives (1). Next, we have to think about it and reflect: do we really need and want this? Based on needs and values, do we judge either positively or negatively? Can the technology be adapted and respond to different needs from different stakeholders? In such a discussion (2), different stakeholders are trying to become contemporaries. But this is not meant to remain something for specialists, but it should take all relevant social stakeholders into account. All interests should be represented like at a roundtable. This entails a search for a common support base (3) and, finally, backed by such agreement the innovation can be adopted, and realized or refused, technically and institutionally (4).

Responsible Innovation, understood as a dialogical process as described, typically seems to be a Western tool. It presupposes public debate and therefore a public space and civil society in which many stakeholders participate in an open discussion, either in support of or opposition to the innovation. That leaves the state authorities, ideally speaking, in a secondary role. They come in first by facilitating the debate, and after the societal process of debate and discussion has reached some form of agreement (however preliminary that may be), they may come in a second time by creating a framework of laws and policies to implement or regulate the technology.

For that reason, there is a direct link between Responsible Innovation and inclusive institutions. Responsible Innovation requires an inclusive state, in which all citizens can consider themselves as represented; and also at least to some degree an inclusive and non-extractive economy, by which all citizens are empowered and capable of participating in the public debate. Is that also the state of civil society in the states and among the stakeholders around the Indian Ocean? This is clearly not the case. On the contrary, there is no striving for consent, and every power is pursuing its own policies. There is a massive presence of the US Navy in the Indian Ocean and a strong political and even military presence of the United States in the Middle East after the conflicts in Iraq and Afghanistan. The Iraqi War cost US$800 billion. On the other side, there is a massive policy of building harbours in order to get a foot on the ground and secure its interests by China in countries around the Indian Ocean. China spent an estimated $1000 billion on such efforts. The situation is further complicated by the rivalry between India and China as local superpowers [3]. And finally, it is further complicated by rising nationalism, the Hindu and Islam revival, an increasingly strong grip of the Chinese Communist Party on public opinion and civil society in China, and finally by the ultraconservative policy of 'America first' by the Trump administration. Increasingly, the Old Silk Routes over land and sea are subject to economic and political interests that seem to be in a stalemate [3]. It seems that not only Kaplan, but all stakeholders involved look upon the ongoing competition as conducted by mere forces in space. As mere forces in space, however, they do not have sufficient understanding of their long-term traditions, of the higher moral powers that also drive them and their societies. Can they do better?

2.5.2 Living backwards while moving forwards

We all are thrown forwards in time by the rapid events of globalization. Terrified by the uncertainties of the future, we all the more try to live backwards. The present world leadership mirrors this situation in its political expression and discourse (note: at the moment of writing, 2018). Trump tries to revive the great old America of the 1950s. Erdogan tries to go back to the great Turkey of the 19th century. Putin tries to do the same with Russia. Modi tries to bring India in line with old-time paternalistic/authoritarian Hindu rule. Xi tries to do pretty much the same with China. Europe tries to close itself off from Africa and the Middle East. We, that means all of us, in all regions and states, don't want the others to be too close and we use the public world theatre primarily to assert ourselves.

That means we are in denial. The world is moving forwards in terms of economic and social integration. The more momentum this movement is gaining, the more reluctant we seem to be to participate in it. Rosenstock-Huessy called this phenomenon 'time lived backward' [44]. Thrown into a completely new situation and receiving new opportunities for self-affirmation at the same time (economic flourishing, for instance, in Turkey, India, China), the first reaction is to live out the old dreams of the past, even if these are unrealistic and even if the magnificent past that is revived may never have existed. Only after a while people come to their senses and face the present and its challenges as different from the dreams of the past.

How did the globalized world come to live in such a short time perspective? One of the main products of the Judeo-Christian tradition is the inspiration to change the world by means of technology. If nature, just like rulers, is desacralized, it is open for exploitation and instrumental use. Technical innovation weakened the ties of all countries and regions to their old traditions. It is due to the success of this driving force of Western civilization that we are thrown together in one space while losing our sense of time, past and future, belongingness and novelty [45]. As a result, we do not understand the meaning of those long-term traditions and values any more. We have a petrified one-minute-item type of understanding of those traditions, which turns them into justifications of our self-interest. We turned those values into worldviews instead of living powers of self-restraint. As a consequence, Hindus, Muslims, Buddhists, Christians, Communists and Liberals struggle for their share and feel threatened by what they do not understand in the other 'worldviews.'

There are four laws of technology that make themselves felt and that ask for a deeper understanding in order to meet this situation:

1. The advance of technology enlarges space. Thanks to modern traffic, the internet and so on, we can reach out to distant regions and we are

starting to live in one world society. We are living close by even distant neighbours.
2. The advance of technology speeds up time. Processes that were very slow in the past do not take as much time, and in our biographies we live through many more changes than past generations ever did.
3. The enlargement of space and the speeding up of time together turn us into uprooted individuals. We lose the connections with the communities in which we originated, and we are uprooted from our traditions. The age of technology, by turning us into daily producers and consumers, shortens our time perspective to our daily affairs.
4. All of us have to learn the art of establishing new communities and leaving them again when the time comes. These changes of living from one to the other community, however, need to be compensated by living more intensely as part of such a community during a limited time. Instead of being rooted once and for all, we need to learn the art to take roots time and again, to move from one value to another value timely and temporarily.

In order to live in a global society, human beings have to turn to a 'planetary' form of living. In the old empires, the stars were fixed at the sky and traditions and habits were as fixed as the stars. Only the planets were on the move; that is also the meaning of the word planet, 'plano,' to roam. In unexpected circles, they moved along the sky. They were the exception. This is a metaphor that can serve well to illustrate our existence as individuals in a global society. We have access to all times and all people, but paradoxically, a lack of roots. Concretely that means that in order to compensate for our uprootedness we have to inherit the values of quite different civilizations and integrate them into our personal existence and social relationships in unexpected circles.

How would that work regarding these two or four streams of tradition that meet in the Indian Ocean and compete as forces in space? The small space of a globalized world 'forces' these power blocs and societal traditions to open up and speak to each other, lest they destroy each other. The shouting between Kim Jong Un from North Korea and Donald Trump from the United States may be paradigmatic for what might happen. Their less well-understood self-interest pushes them to oppose and fight each other. But they can't fight, lest they destroy each other. That turned their shouting into empty words, to such an extent that the world was not holding its breath in this conflict. Nobody believed they were serious, actually. Maybe they, confronted with the violence they could unleash, were re-minded to self-restraint. Some values from the past they might have forgotten while shouting re-entered their minds. Re-minded they were of their respective traditions of self-restraint, which are in line with a more reflective and inclusive understanding of their self-interest.

Our spiritual value traditions turn the power game into a dialogue by forcing the question upon us: where do we come from? Where do we go? From where did we originate? Where is our pluralist, yet common, destination? And considerations of this kind can take the seriousness away from the power game and turn that dangerous word into what it really should be and mostly is not: a game. Our value traditions inform us with a broader repertoire of options to deal with each other, respect each other and yet keep each other in check, even if only by well-understood self-interest – but well understood! [15]. The stakeholders in this game might secure more of their self-interests inasmuch as they are able to consider the moral drivers of the other stakeholders involved. The recent plans from the United States and many of its allies to contribute to the One Belt, One Road initiative and complement several of the plans of China show how the game can also be played. That may set a different tone of responsiveness to each other. If taken and appreciated in this way, the 'higher grammar' [4] of responsiveness to the value traditions embodied by the different partners will also take the Responsible Innovation approach to a higher level. Responsible Innovation will become more than an exchange of interests between competitors who merely understand themselves and each other as competing forces. It will become more than a mere 'tool' in the negotiations. It will help to not only negotiate but to talk to real human beings with a storyline of their own, a storyline that calls for respect and continuation and translation of their respective values into a common future. Our respective value traditions give us a face and humanize the debate. They coordinate us towards each other. This responsive coordination towards each other needs to be established before the information can flow.

2.6 Conclusion

This chapter is meant to find a moral-spiritual methodology to avoid a clash of cultural values and institutions, called by Huntington the clash of civilizations, along the One Belt, One Road project in the Indian Ocean. To that purpose, I developed the concept of value-sets, which may contain and establish an equilibrium between different and contradictory values, even within one cultural tradition. Where different cultural traditions meet, this process of integrating different value-sets and value-orders or sequences becomes even more complicated and intricate. This is usually called the field of intercultural management, although the term 'management' is not well in place, because it assumes that we are above the situation and can handle (manage = manus = hand) it. That is not correct. We are part of it, that is, part of the living values that we represent, and we cannot avoid expressing them ourselves from the moment we start talking. That means we cannot bypass our subjectivity. As participants in the process, we are all biased, partial, partisans. But being so we can become partners in the process! As part of the situation, we cannot do otherwise but be what we are, and at the same time, be open for what we are not and what others represent and express in their ways of life, values and institutions. This situation calls for self-restraint, and for that reason I investigated four traditions of self-restraint within the different streams of culture and civilization.

While each civilization participates in the usual repertoire of family/clan and hierarchy/state institutions in East and West, four key streams of countervailing values and critical traditions can be distinguished, which are not only critical but also a condition for larger-scale rule and social organization. They are traditions of denial and self-denial or self-restraint, which are identified as an indispensable ingredient of larger-scale rule in all cultural traditions. Buddhism reduces all will to be by the power of the eye to nothingness, Maya. Taoism reduces all claims for recognition and status to nothingness by not giving a damn, of Wu Wei. Judaism reduces all powers that be to nothingness by claiming justice beyond all legitimating forces (gods), making the peoples of the earth drops in the ocean. Christianity reduces all powers of the present to nothingness by sacrificing success in the present to fruitfulness in the future, opening a road for revolutionary/evolutionary change. This future-directed force of Christianity or, if one prefers, the Western world integrated all cultures and political entities into one globalized world, turning all of them into neighbours, living just around the corner. This movement paradoxically leads to a new and unexpected imperative, the next step into the future, which precisely consists in integrating the heritage of the past into this future-directed power, into one globalized world. The world wars of the 20th century urged that imperative on all people of this planet, that is, to open up and talk to each other, learn from each other's different value traditions, lest the world goes to pieces.

If societies open up and speak to each other, their value traditions may not only exclude each other (sometimes) but more often support and correct each other. Buddhism and Christianity, as an example, influenced each other a long time ago. Via stoicism (Zeno) and later the Christian monasteries, the Buddhist attitude of withdrawing from worldly affairs, turning inside and looking upon the world as a screen without attachment (Maya), became a constitutive part of the Christian tradition. And, in turn, the Christian notion of compassion with human suffering found wide access in the Buddhist tradition and took this tradition more into the orbit of social responsibility than otherwise might have happened. The same counts for Hinduism and its understanding of the caste system, roles of males and females and professions. Slavery in the West was a black page in its history – the Emancipation Reform was first in Russia in 1861 and then slavery was abolished in the West in 1863, while all other forms of slavery in the Eastern empires came to an end. No sane Muslim or Hindu would in our days be inclined to claim that their religion justifies slavery any more.

Many Westerners who live under the stress of large-scale modern organizations find relief in the self-effacement of Taoism and/or in the self-annihilation of Buddhism and Hinduism. This mutual penetration of traditions is taking place before our eyes.

There is one condition for this process of mutual interpenetration of value-sets to have its course in the case of competition and conflict along the One Belt, One Road project in the Indian Ocean. The different partners and parties competing with each other have to exercise the capability of not only living in their own time but also drawing from older value layers – the meeting between Kim Jong Un and Donald Trump would not have become so grotesque. Both in the history of East and West there are values of self-restraint and self-effacement, either of Christian or Buddhist or Taoist origin. It is true that the law of technology makes us live in a short time perspective of mere space, in the cycle of production and consumption of daily life. But it appears to be more urgent to exercise the capability to live not only in the moment, but to receive and remember the past and prepare for the future. We should live at least in three times: past, future, present [46]. The exercise of taking our different past histories and our shared responsibility for the future into account should turn us into participants in one human endeavour. That is what I have put forward in this chapter. With the resources from the past, and an open eye for the future, we can become contemporaries in the present.

For the concept of Responsible Innovation this means that it may not be possible to deal with it only in the Western way of a combination of public debate and state regulation. Such an approach can only work inasmuch as the other stakeholders already have adopted Western-type states and civil societies. I already pointed out that such a process is taking place, but for one thing, the process is still underway, and for another thing, we cannot expect non-Western traditions to get to precisely the same result as the West [1, 47]. These traditions might find different solutions for having a strong, inclusive state which is at the same time accountable to an open civil society [30]. Family, clan, tribe and caste may survive for a long time to come in the South and the East, and at the same time, we see traces of revival of closed in-group mentalities [48, 49] in the West. Future solutions might look somewhat like Indian democracy, in which a centralized parliamentary system also integrates vertical networks of caste and region, combining a national Indian identity with more distinct regional and/or group identities represented in federal and regional parliaments [47].

Whatever institutions will be invented, designed and implemented, they will have to consider the four countervailing traditions and voices described in this chapter. Either as part of a religious or as part of a more secular ethical discourse, these moral-spiritual countervailing forces need to become part of the 'system.' They call forth and reinforce each other. Taking them into account coordinates the stakeholders in the Responsible Innovation process towards each other. Thereby they take the Responsible Innovation approach above and beyond the usual dealings of self-interested parties understood as mere forces in space. The Responsible Innovation process between partners/parties that voice long-lasting traditions can make the Responsible Innovation approach more inclusive and comprehensive. They do not need to understand each other only as competing forces, to be calculated within a zero-sum power game in the globalized space of one world society. They can also understand each other as harbouring important value traditions, as partners living in one house, with many rooms and open doors. The preservation of what is most precious in their traditions promises each of them a room in the one house of humanity they are building.

References

[1] Rosenstock-Huessy, E., *Out of Revolution: Autobiography of Western Man*. Argo Books, New York (original 1938), 1993.

[2] Huntington, S.P., *The Clash of Civilizations and the Remaking of World Order*, Simon & Schuster, New York, 1996.

[3] Kaplan, R.D., *Monsoon: The Indian Ocean and the Future of American Power*, Random House Publishing Group, New York, 2010.

[4] Rosenstock-Huessy, E., *Practical Knowledge of the Soul*, translated by M. Huessy and F. Moltke, Wipf & Stock, Eugene, OR, 2015.

[5] Mahbubani, K., *The New Asian Hemisphere*, Perseus, New York, 2008.

[6] Joas, H., *Die Entstehung der Werte*, Suhrkamp, Frankfurt, 1999.

[7] McIntyre, A., *After Virtue*, University of Notre Dame Press, Notre Dame, IN, 1984.

[8] McCloskey, D.N., *The Bourgeois Virtues: Ethics for an Age of Commerce*, University of Chicago Press, Chicago, London, 2006.

[9] Rosenstock-Huessy, E., *The Multiformity of Man*, Argo Books, New York, 1973.

[10] Waltzer, M., *Spheres of Justice: A Defense of Pluralism and Equality*, Basic Books, New York, 1983.

[11] Voegelin, E., *The New Science of Politics: An Introduction*, The University of Chicago Press, Chicago, London, 1952.

[12] Rosenstock-Huessy, E., *Soziologie II*, Kohlhammer, Stuttgart, 1958.

[13] Ferry, L. & Gauchet, M., *Le Religieux après la Religion*, Grasset & Fasquelle, Paris, 2004.

[14] Girard, R., *Des Choses Cachées depuis la Fondation du Monde*, Grasset, Paris, 1978.

[15] Morgan, G., Selfishness and Unselfishness, *The Journal of Philosophy*, 37(15), pp. 401–407, 1940.

[16] Zimmer, H.R., *Myths and Symbols in Indian Art and Civilization*, Campbell, J. (ed.), Princeton University Press, Princeton, NJ, 1943.

[17] Chen, E.M., *The Tao Te Ching: A New Translation with Commentary*, Paragon House, St Paul, MN, 1989.

[18] Rosenstock-Huessy, E., *The Driving Power of Western Civilization*, The Beacon Press, Boston, MA, 1950.

[19] Armstrong, K., *A History of God: From Abraham to the Present, the 4000 Year Quest for God*, Heinemann, London, 1993.

[20] Rosenstock-Huessy, E., *Speech and Reality*, Argo Books, Norwich, 1970.

[21] Eisenstadt, S.N., The Axial Age Breakthroughs. In: Eisenstadt, S.N. (ed.), *The Origins and Diversity of Axial Age Civilizations*, State University of New York Press, New York, pp. 1–29, 1986.

[22] Taylor, C., What Was the Axial Revolution? In: Bellah, R.N. & Joas, H. (eds.), *The Axial Age and Its Consequences*, Harvard University Press, Cambridge, MA, London, 30–46, 2012.

[23] Frankopan, P., *The Silk Roads: A New History of the World*, Bloomsbury, London, 2015.

[24] Acemoglu, D. & Robinson, J.A., *Why Nations Fail? The Origins of Power, Prosperity and Poverty*, Profile Books, London, 2013.

[25] Hsu, C.Y., Historical Conditions of the Emergence and Crystallization of the Confucian System. In: Eisenstadt, S.N. (ed.), *The Origins and Diversity of Axial Age Civilizations*, State University of New York Press, New York, pp. 306–325, 1986.

[26] Buber, M., Die Lehre vom Tao. In: *Martin Buber, Werke*, Bd. I, Kösel Verlag, Lambert Schneider, Heidelberg, pp. 1021–1053, 1962.

[27] Julien, F., *Detour and Access: Strategies of Meaning in China and Greece*, Zone Books, New York, 2000.

[28] Elman, A.B., Changes in Confucian Civil Service Examinations from the Ming to the Ching Dynasty. In: Elman, B.A. & Woodside, A. (eds.), *Education and Society in Late Imperial China, 1600-1900*, University of California Press, London, pp. 111–149, 1994.

[29] Dunbar, G., *A History of India: From the Earliest Times to 1939*, Vol. I, Nicholson & Watson, London, 1949.

[30] Fukuyama, F., *The Origins of Political Order*, Exmouth House, London, 2011.

[31] Huxley, A., Introduction, *The Song of God, Bhagavad-Gita*, translated by S. Prabhavananda & C. Isherwood, Mentor, New York, 1944.

[32] Shankkar, A., *Accidental India: History of the Nation's Passage through Crisis and Change*, Rupa Publications, New Delhi, 2013.

[33] Rosenstock-Huessy, E., *The Christian Future, or the Modern Mind Outrun*, Charles Scribner's Sons, New York, 1946.

[34] Voegelin, E., *Order and History*, Volume 14, Hogan, M.P. (ed.), *Israel and Revelation*, University of Missouri Press, Columbia, MO and London, 2001.

[35] Picht, G., *Hier und Jetzt: Philosophieren nach Auschwitz und Hiroshima*, Bd. I., Klett-Cotta, Stuttgart, 1980.

[36] Berman, H.J., *Law and Revolution: The Formation of the Western Legal Tradition*, Vol. I, Harvard College, Cambridge, MA, London, 1983.

[37] Braudel, F., *The Perspective of the World: Civilization and Capitalism 15th to 18th Century*, Vol. 3, Harper & Row, London, 1984.

[38] Gupta, A., *Red Tape: Bureaucracy, Structural Violence of Poverty in India*, Duke University Press, Durham, NC and London, 2012.

[39] Stackhouse, M.L., *Creeds, Society and Human Rights: A Study in Three Cultures*. Eerdmans, Grand Rapids, MI., 1984.

[40] Landes D.S., *The Wealth and Poverty of Nations: Why Are Some So Rich and Others So Poor*, WW Norton, New York, 1998.

[41] Collins R., *The Sociology of Philosophies: A Global Theory of Intellectual Change*, Harvard University Press, Cambridge, MA, London, 1998.

[42] Chang, W., Legal Education in Ch'ing China. In: Elman, B.A. & Woodside, A. (eds.), *Education and Society in Late Imperial China, 1600–1900*, University of California Press, London, pp. 292–339, 1994.

[43] Stilgoe J., Owen R. & Macnaghten P., Developing a framework for responsible innovation. *Research Policy*, 42(9), pp. 1568–1580, 2013.

[44] Rosenstock-Huessy, E., *Die Sprache des Menschengeschlechts*, 2 vols, Verlag Lambert Schneider, Heidelberg, 1963.

[45] Rosenstock-Huessy, E., *Friedensbedingungen der planetarischen Gesellschaft: Zur Ökonomie der Zeit*, Ed. Rudolf Hermeier, Agenda Verlag, 2001.

[46] Kroesen, J.O., *Planetary Responsibilities: An Ethics of Timing*, Wipf & Stock, Eugene, OR, 2014.

[47] Kaviraj, S., Modernity and Politics in India. In: Eisenstadt, S.N. (ed.), *Multiple Modernities*, Transaction Publishers, New Brunswick, NJ, pp. 136–162, 2002.

[48] Hyden, G., *African Politics in Comparative Perspective*. Cambridge University Press, Cambridge, 2013.

[49] Bauman, Z., *Postmodern Ethics*, Blackwell, Cambridge, MA, 1993.

[50] NASA, Gwadar satellite view showing island created by 2013 earthquake.jpg, https://earthobservatory.nasa.gov/images/82146/earthquake-births-new-island-off-pakistan, accessed on: 6 December 2018.

[51] Michalakeas & Garay Garcia, this volume, Chapter 6.

CHAPTER 3

Opening up to the world: Chinese foreign policies as a framework for indigenous development along the New Silk Road

Yan Zhang[1,2] & Wim Ravesteijn[2,3]

[1]School of Management, Harbin Normal University, China
[2]Department of Values, Technology and Management, Delft University of Technology, The Netherlands
[3]Beijing Information Science and Technology University, China

Abstract

As the Chinese have entered the landscape of global relations and governance, the new foreign policies under Xi Jinping have been watched and analysed by many governments and researchers. Not only from the viewpoint of Chinese history but also from an international development perspective, the new opening-up policy towards the world in China illustrates the new goals of comprehensive improvement and national rejuvenation. The Belt and Road Initiative (BRI) is a major project for promoting openness as well as for intensifying cooperation with other countries along the land and sea routes involved in the Initiative. The purpose of this chapter is to look at the history and main characteristics of the present opening-up policy, while also demonstrating the indigenous opportunities and effects for some countries along the New Silk Road. Ultimately, it can be concluded that the BRI provides possibilities for contributions to regional governance and international development as well as indigenous growth, although some challenges and risks should not be overlooked.

Keywords: Opening-up of China, foreign policy, BRI, Silk Road, cooperation, development

3.1 Introduction

During the 19th National Congress of the Communist Party of China (CPC), on 18 October 2017, President Xi Jinping presented an important report in which he expressed his wish to make new progress in making China, on all fronts, increasingly open to the outside world [1]. This aspiration of completely opening its borders illustrates the emergence and impact of China as a great power in the world under President Xi's administration. It is the continuation and sharpening of a radical transformation of earlier policies. In this respect, the question may arise: why this change in attitude towards foreign countries?

Historically, China has been an open country, from the flourishing Tang dynasty to the early Ming dynasty, which culminated in the great Zheng He's journeys at the beginning of the 15th century but changed afterwards. In the following four centuries, China fell into complete self-seclusion until the Opium Wars forced it to re-open its borders again, which also introduced China into a long 'Century of Humiliation.' After the foundation of the People's Republic of China (PRC), its leader Mao Zedong took a strong outwards orientation in the sense of taking joint actions with the Soviet Union in order to learn from that country (xiangsu yibiandao, 向苏–边倒), but this relationship broke in 1960. After that, China was completely re-closed to the world for almost 20 years. When Deng Xiaoping became the real leader of China, he proposed the well-known reforms and opening-up policies from 1978, originally setting up four Special Economic Zones (SEZs). From then on, economic growth has become the main focus, which essentially introduced the new Chinese development model.

Consequently, China has become the world's second-largest economy with a GDP of 82 trillion yuan in 2017. China's real GDP has grown with an average of 9.6% per year in the period from 1980 to 2015, while the GDP per capita increased from US$200 in 1980 to over US$800 in 2015 [2]. Generally, besides economic growth, which resulted in better living standards, the Chinese government, people, business corporations and institutions indisputably made progress in other domains, including political stability and environmental quality.

The 30 years of experiences with the outside world have turned out successfully and have supported the Chinese leadership in widening and deepening its view of the global community and its changes. Xi's rhetoric of the Chinese dream and his policies urging further opening-up imply that China is no longer a weak force in the world and new plans have been drawn for a global profile.

Things have culminated in the current ambition of 'a moderately prosperous society in all aspects' (quanmian xiaokang society, 全面小康社会) in order to realize the Chinese dream of national rejuvenation [1]. Xi and his leadership promote 'a two-stage development plan' from 2020 to the middle of the 21st century to achieve the 'two centenary goals' (liangge yibainian, 两个–百年), one of which is achieving the 'quanmian xiaokang society' goal, while the other is building a modern socialist country [1].

In realizing these aims, further opening-up is necessary for the new Chinese plans. China has grown to be a country of non-negligible character on the global stage and is renewing its position closer to the centre of the world. Xi has frequently emphasized that 'China will not close its door to the world,' which sustains the opening-up policy, as China's economic reforms and successes provide the Chinese government with more courage and capabilities to directly face up to and participate in global competition. For example, Chinese foreign exchange reserves reached almost US$4 trillion at the end of June 2014, while the amount was only US$212 billion in 2001. Another important transition can be supposed to account for this success. Chinese mergers and acquisitions (M&A) spent US$227 billion abroad, which was six times more than the spend of foreign companies on acquiring Chinese corporations [3].

Simultaneously, in line with the new opening-up policy, the 'One Belt, One Road' (OBOR) plan was issued in 2013, constituting a priority and a new decision-making mechanism for the global political economy, which is well known as the 'New Silk Road' in distinction from the 'Old Silk Road.' It aims to increase openness, cooperation and linkages with countries eastwards, southwards and westwards along land and sea routes. It is officially named the 'Belt and Road Initiative' (BRI) [4, 5], containing the 'Silk Road Economic Belt' (SREB) and the '21st Century Maritime Silk Road Initiative' (MSRI) respectively. The BRI aims to promote bilateral and multilateral partnerships in Asia, consolidate China's leadership status in the region and strengthen China's continued efforts in the field of national rejuvenation as a 'Comprehensive National Power' (see Figure 3.1).

In addition, there are also other reasons for the continuation of the open-door policies. The main one is that China, established and perceived by its leadership as a global leader [6], takes up its corresponding responsibilities. Xi has constantly declared that China will contribute to building 'a community of common destiny' (mingyun gongtongti, 命运共同体) for the whole of humankind. This new and crucial concept was embraced at the 55th UN Commission for Social Development in February 2017 [7] with regard to African development. It was the first time the UN quoted China's significant mission for the world [7]. While the United States withdrew from the Paris Agreement in 2017 [6], Xi and the Chinese government strongly declared that China would persist in taking up responsibility for global climate change action and would implement the commitments from the Paris Agreement [8]. Consequently, China has become a regional and international leader in various respects.

This chapter discusses Xi's new foreign policy, from both narrow (1949–present) and broad (third century BCE–present) historical perspectives and the way it assumed shape in the course of time (first four SEZs, after that hosting the World Trade Organization [WTO] and the Olympic Games, and then APEC 2014, G20 2016, Shanghai Free Trade Zone, the BRI, etc.). It also investigates what this new policy means for the countries of the New Silk Road: its impacts and the possibilities it provides for indigenous development. It will be clear that China thinks to profit from these new measures, but what about other

Figure 3.1: Belt and Road International Cooperation. (Source: Wikimedia, 2018 [65].)

countries? China aspires to stimulate the economic development of associated countries, but are these ambitions realistic? And under what conditions can other countries profit from China's actions?

This chapter aims to contribute to the assessment of and the discussion about the effects of the New Silk Road programme for African, ASEAN (Association of Southeast Asian Nations) and European countries. After this introduction providing the background to the research, we proceed in six sections. In Section 3.2, a broad historical overview is given showing the Chinese moving between opening borders and closing them again. The third section focuses on the principles of Chinese foreign policy since the reform and opening-up programmes in the period from 1978 to 2012. The characteristics of the new opening-up efforts under Xi Jinping are discussed in Section 3.4. The fifth and sixth sections highlight the New Silk Road initiative and its implications for other countries along the BRI. The discussion and conclusion emphasize the responsibilities and collaboration involved in the implementation of China's foreign policies and development strategy.

3.2 Historical overview

When discussing China's foreign policy, it is extremely interesting to review the Chinese opening and closing of its borders from the Qin and Han period to the Qing dynasty. The famous Southsea Path to develop overseas trade can be traced back to the Qin dynasty [9]. During the Han dynasty, the Old Silk Road emerged as a trade route between Europe, India, China and many other countries on the Afro-Eurasian landmass. When the Tang dynasty ruled, the heyday of Chinese feudal society, the country implemented an opening-up policy to the world, not only for economic trade but also for political and cultural communication. During the following Song, Yuan and early Ming dynasties, some openness continued. For instance, Zheng He made seven sea voyages to more than 30 countries in the Western Pacific and Indian Ocean area in the 1405–1433 period, which constitute well-known events in maritime development history [10]. But China switched to a closed-door policy in order to protect imperial authority from the mid-Ming dynasty until the Qing dynasty. China, however, was forced to open its borders to the Western powers after the Opium War of the 1840s and experienced a prolonged period of domination and instability as well as decline over the following 100 years.

Since Mao Zedong and the CPC took the lead in building a 'new China' – the PRC – in 1949, independence and territorial integrity have been confirmed and preserved in China [11]. In the first Five-Year Plan of China, Mao allowed a limited openness to international investments and technological cooperation with Western countries [9], which, unfortunately, was not accepted by many countries. After that, the Chinese leadership was especially open to the Soviet Union, studying everything from the so-called 'respectable eldest son' (Laodage, 老大哥) [12]. Generally, Mao's policy towards foreign relationships and diplomacy was only focused on socialist countries. However, because of the dominant position of the Soviet Union and the stress when it suddenly withdrew its experts from China in 1960, cancelling 257 projects [13], China returned to closing. The Chinese kept a closed door to the world for the next 20 years until Deng Xiaoping proposed reform and the opening-up of foreign policy in 1978.

3.3 Principles of China's foreign policy under Deng Xiaoping

After suffering the ten-year 'Cultural Revolution,' the Chinese government realized it was of the utmost importance for China to restore a normal economic order and the quality of people's lives in general. Consequently, Deng Xiaoping initiated his well-known policy of reform and opening in 1978, accepted by the Third Plenary Session of the 11th CPC Central Committee. Meanwhile, Deng emphasized that China should 'Bide its time, hide its brightness, not seek leadership, but do some things' (taoguang yanghui, 韬光养晦) [14–16].

In the following 30 years, China largely stuck to Deng's diplomatic policy of peace and development, steadily keeping to its economic development and national modernization plan. Thanks to reform and the opening-up foreign policy, China obtained tremendous and unbelievable achievements not only in the economic field but also in its political and military position. Especially after joining the WTO in 2001, China realized an economic take-off and a strengthening of its international status [17]. Therefore, from Deng's time until Hu Jintao, the fourth leader of China, Chinese foreign policy insisted on Deng's call and slogan. Two stages can be distinguished: first, the 'bringing in,' second, the 'sending out' [17].

3.3.1 'Bringing in' period: 1978–2000

After the confirmation of reform and the opening-up policy, China successfully started to set up a socialist market economic system, while the Third Plenary Session of the 14th CPC Central Committee proposed the development of an open economy. In order to let in foreign companies, in 1980, the Chinese leadership defined four maritime cities (Shenzhen, Zhuhai, Shantou and Xiamen) as SEZs. From then on, the open foreign policy with Chinese characteristics became the basic state policy in China.

In this period, Chinese foreign business brought in several advantageous resources in various fields such as foreign capital, technology, talents, management experiences and so on. These 'bring in' strategies enabled China to jump ahead regarding capital investment, trade and other economic respects. For example, by 2000, the total trade value of Chinese exports and imports ranked 7th in the world, occupying 7.3% of global trade, according to IMF statistics [17]. And with the Jiang Zemin administration, the nation achieved the goal of being 'moderately prosperous' (xiaokang, 小康), realizing an economic take-off and a strong position in the world market.

3.3.2 'Sending out' period: 2001–2012

The progress in foreign openness was marked by China becoming a member of the WTO on 11 December 2001, which drove the Chinese transition from a large opening-up nation to a strong one. With the rise of modern industry, Chinese enterprises had enough capability to face the opportunities and threats of foreign markets and businesses, becoming indispensable partners on the global stage of trade and competition. According to the '2012 World Investment Report' issued by the UN, direct Chinese foreign investment reached US$74.65 billion, accounting for approximately 5% of the global flow [17], ranking in the top ten in the world. Meanwhile, Chinese foreign exchange reserves swelled to almost US$3.31 trillion in the same year [17].

Simultaneously, China's contributions on the international stage and the responsibility it took in world affairs in general, especially as a 'soft power,' began to be recognized after Hu Jintao stepped down as the general secretary of the CPC. The 2008 Olympic Games epitomized the success of China's foreign openness, radiating the image of a well-off, civilized, inclusive and modernized nation. Subsequently, in 2010, Shanghai successfully hosted the World Expo, which was the most expensive in the history of world's fairs, costing almost RMB 11.964 billion (US$1.8 billion), as well as having the most participants, consisting of 246 countries and international organizations [18]. China, once more, demonstrated its international position as the 'major nation.'

The 'sending out' has been viewed as a great opening strategy towards the global economy, delivering better chances for broader international cooperation and showing that China has obtained unbelievable success in a great many respects.

Looking back at more than 30 years of Chinese foreign policies, the Chinese leadership has strongly embraced Deng's approach, both ensuring steady domestic economic growth and participation and peace abroad. China has clearly benefited from the policy of reform and openness, which contributed to realizing the goal of building 'a well-off society in an all around way' [19]. Furthermore, this foreign policy can be described as a clear programme for China: openness created more opportunities for domestic development and promoted capabilities, while it upgraded national power. Thus, the opening-up policy should be continued by the leadership, providing more active and efficient methods for new developments.

3.4 Characteristics of the opening-up policy

Although China rapidly re-emerged as a major player on the global stage [3], initially, Chinese diplomacy was more passive regarding regional development and international responsibilities [2, 3] because the Chinese leadership followed Deng Xiaoping's principles of 'hiding and biding time' on the international stage [14–16]. Only with Xi Jinping, who was appointed as general secretary of the CPC in November 2012 and as state president in March 2013 [20], did Chinese foreign policy transform itself from 'hiding and biding time' towards a new vision.

This can be traced back to the 'China Dream' (Zhongguo meng, 中国梦) announced by President Xi at the 18th National Congress of the CPC in November 2012, when he referred to 'a moderately prosperous society in all respects' [21]. This phrase returns in many high-level speeches and reports by Xi not only on domestic platforms but also on various international occasions, reflecting China's integration into the international economy and its politics. From then on, Chinese policymakers rolled out a new discourse to the world through notions such as 'China Dream,' 'a community of common destiny' (mingyun gongtongti, 命运共同体), 'one home for all of mankind' (renlei gongtong jiayuan, 人类共同家园) and so on. Together, the changes in the Xi era have resulted in a more proactive, assertive, participant and a more responsible open-door policy [15, 21], which is detailed in the following sections.

3.4.1 To establish a large national foreign strategy: From passive to active

During Xi's initial diplomatic activity as China's top leader in late 2012 [22], he did not hesitate to emphasize that 'the world has increasingly grown into a community where one's destiny is interwoven with that of another,' so China should lead to developing 'a community of common destiny.' The concept continually emerged in Xi's speeches in the following years. Taken together, the Chinese government have proposed and implemented a series of measures beneficial for internal development and their external position in order to achieve the goals of the 'great rejuvenation of the Chinese nation' (zhonghuaminzu weida fuxing, 中华民族伟大复兴) and 'responsible great power' (fuzeren daguo, 负责任大国).

It consists of, but is not restricted to, the assertion of peace with a reduction in the number of troops, ensuring political stability with a national security system, eco-culture construction with carbon emission reduction and smog governance and the improvement of soft power through a new win–win and cooperation diplomatic strategy. In 2015, for example, the Premier Li Keqiang issued a great ten-year national plan – 'Made-in-China 2025' – in order for China to be a world-class industrialized power [23]. In addition, Beijing has spent 36.60 billion yuan [24] from 2014 to 2016 to decrease smog pollution and improve air quality. Consequently, total carbon emissions in China have substantially been reduced until 2016. It is clear that the new Chinese leadership is actively seeking a new increase in China's national economic and political power.

3.4.2 To participate in global governance: From peripheral to central

In line with Xi's new foreign policy, Chinese policymakers understand it is time for China to make increased efforts on behalf of the global economy. In the first year of Xi's presidency, he indicated several times that the door of China would continue to be open to the world in a profound way and he promoted inclusiveness along with openness.

Since the 18th Party Congress in 2012, China's influence in the G20 group of most developed countries has grown and the government has succeeded in compliance with international rules in various domains, such as energy, food security, climate change, anti-corruption, infrastructure and so on [21]. In September 2016, it was considered a valuable opportunity for China to play a key role in Sustainable Development worldwide, Xi saying that 'the first action plan has been formulated for implementing the 2030 Agenda for Sustainable Development, and for the first time, cooperation is being carried out to support African countries and LDCs in their industrialization.' Also, the World Economic Forum (WEF) has become one of the main interests of the Chinese government and business. Xi made a four-day visit to Switzerland, visiting the 2017 WEF annual meeting and giving two significant speeches, showing Chinese wisdom, confidence and responsibility in the face of the global economy and world development [25]. He pointed out that China is a contributor to economic globalization and its door will be kept wide open while featuring win–win cooperation as well as fair and equitable governance [25]. During this forum, China showed the importance it attaches to the world economy by strengthening innovation and cooperation with European countries as a national strategy. China's determination within global governance is obvious.

3.4.3 To improve influence as a regional leader: From tributary to decision-making

In addition to its efforts on behalf of global governance, China has undertaken great efforts towards the creation of safer and sustainable regional development relationships with countries in the Asia-Pacific region, Africa and Europe.

First, the Asia-Pacific Economic Cooperation (APEC) is a leading economic organization for Asia-Pacific countries. Chinese leaders continued to participate in the APEC CEO Summit Meeting, delivering keynote speeches. Significantly, China successfully showed its 'home field' achievements and it was sincere in discussing issues of regional economic integration and economic innovation development as well as the promotion of connectivity infrastructure during the 2014 22nd APEC Economic Leaders' Meeting in Beijing. In November 2017, Xi Jinping gave a keynote address at the APEC CEO Summit, stating that 'Over the past four years, China's economy has grown by 7.2% on the average annually, contributing over 30% of global growth. China is now the main driver powering global growth.' In Xi's vision, China will intensify regional economic integration and strengthen bonds with other countries, ensuring and guiding development, peace and stability in the Asia-Pacific region and beyond.

Another important activity for the Chinese leadership that should not be overlooked is its participation in the BRICS (Brazil, Russia, India, China and South Africa) group. China, as the initiating member state of the five major emerging national economies, helped to establish the New Development Bank (NDB) in 2015 and successfully organized the 9th BRICS Leader Meeting and the BRICS Business Forum in Xiamen in 2017. According to Xi, China is a contributor and leader in the BRICS group, which will use regional strengths and influences to deliver more benefits to more people. Furthermore, following in the open spirit, the Shanghai Free Trade Zone (FTZ) was established in 2013 and the Central Committee accelerated the implementation of the programme and established another seven FTZs before 2016, which now constitutes its primary free trade strategy [26].

In conclusion, China has redefined its national development strategy to not only include domestic development but foreign policies as well. Xi's new slogans, marking a proactive and influential leader, provide China with a historically new role on the world stage.

3.5 The New Silk Road initiative

The New Silk Road has attracted worldwide attention since it was raised by Chinese President Xi Jinping in the autumn of 2013, especially among Asian, African and European countries. During his visit to Kazakhstan and Russia, President Xi announced the concept to build the 'Silk Road Economic Belt' (SREB) as the first step to connect China to Europe via Central Asia [20, 27]. Xi proposed to jointly construct the MSRI in the following Indonesia trip in order to not only develop China's economic ties along the sea route [28] but also enhance China's new opening-up foreign policy and its international contributions. The Belt and Road Initiative (BRI), as it was officially designated in 2015, has become the defining Chinese foreign strategy in the new era and a field of opportunity for all the involved countries and institutions [29].

In May 2017, China organized and held the momentous international summit, the Belt and Road Forum (BRF). During the opening ceremony of the BRF for International Cooperation, Xi delivered the influential speech 'Work Together to Build the Silk Road Economic Belt and the 21st Century Maritime Silk Road,' calling for the building of a new road connecting different civilizations devoted to peace, prosperity, innovation and openness. In order to fully achieve the BRI, many plans, standards and implementations have been concluded by the Chinese government, corporations and organizations, and key characteristics can be summarized in terms of financial and regional cooperation as well as by plans.

3.5.1 Financial and regional cooperation

A primary and significant action for the BRI to come to fruition was the foundation of the 'Silk Road Fund' (SRF) at the end of 2014, the initial capital of which amounted to US$40 billion, but which was later increased. The SRF is an infrastructure financing vehicle to fund and support multilateral connectivity and economic trade cooperation along the New Silk Road. According to the leader of SRF, Yang Luo, the SRF can support the sustainable development of middle-term or long-term infrastructure in the involved countries, and 17 project contracts have been signed covering US$80 billion of the total investments in the period until December 2017 [30]. Meanwhile, during the 2017 BRF meeting, Xi proclaimed that China will add another 100 billion yuan to the SRF for contributions to

BRI constructions. The SRF will also strengthen advantages as to equity investment and resource development as well as capacity and finance cooperation.

With regard to the economic goals and its strategic position, China decided to launch the Asian Infrastructure Investment Bank (AIIB). Xi Jinping met the heads of delegations from the 21 original founding member countries in 2015 and indicated the importance of the AIIB for Asian future development [31]. Synchronously, Xi proposed the idea of establishing the AIIB in a broader circle at the 2015 Boao Forum at the Asia Annual Conference, emphasizing the infrastructure of Asia would 'advance complementary and coordinated development' [21]. Consequently, the AIIB was founded on 25 December 2015 and opened on 16 January 2016. The AIIB has grown to 84 approved members from around the world. The goals and principles of the AIIB are to 'better connect people, services and markets that over time will impact the lives of billions and build a better future' [32]. The AIIB is a regional multilateral developed institution of Asian governments and the first multilateral financial organization proposed by China.

As the largest investor and shareholder in the AIIB, China will contribute US$29.78 billion to hold more than a 26% share, according to the AIIB Articles of Agreement (AoA) [21, 33]. Through the AIIB, China became the actual leader in regional development, especially in the financial respect, simultaneously setting a new standard for international economic governance. At the end of 2017, the AIIB provided the resources for investments into 24 projects among Asian countries, covering loans totalling US$4.2 billion, mainly in the fields of energy, transport and urban infrastructure. More importantly, the AIIB, as a significant milestone of the BRI, is becoming the tool for Beijing to connect the Chinese government with other countries along the New Silk Road. AIIB will provide China and other BRI countries with more resources, more opportunities and more advantages.

3.5.2 Top-level official plans

In 2015, China officially issued 'The Vision and Actions on Jointly Building the Silk Road Economic Belt and the 21st Century Maritime Silk Road,' which suggests 'promoting policy coordination, connectivity of infrastructure and facilities, unimpeded trade, financial integration and people-to-people bonds' (wutong, 五通). It is the first official document with reference to the regional ocean economy and cooperation strategy, which also shows the top-level design and strategic plan for the New Silk Road. In the same year, another plan was released by the Office of the Leading Group for the Belt and Road Initiative, 'Action Plan on Connecting the Belt and Road by Standards (2015–2017),' to indicate the ten aspects aimed at promoting and improving quality and technology with certification, standards, measurement and accreditation in BRI countries [34, 35].

Consequently, two other top-level documents were issued to promote the quality of cooperation among BRI countries in 2017. One was the 'Vision for Maritime Cooperation under the Belt and Road Initiative,' in which the Chinese government advocated the Silk Road Spirit – 'peace and cooperation, openness and inclusiveness, mutual learning and mutual benefit' – and proposed to 'engage in all-dimensional and broad-scoped maritime cooperation and build open and inclusive cooperation platforms' [5]. Since then, some port and marine projects have been carried out for the 'Blue Partnership' [5] along the MSRI to promote the Chinese ocean strategy and international economic cooperation. The other is 'Building the Belt and Road: Concept, Practice and China's Contribution,' in which China intimates to build and open 'five routes,' 'six corridors,' 'six means of communication' and multiple ports and countries for the BRI [35]. Meanwhile, three ocean-linked 'blue economic passages' [5] were outlined for the marine connection and further cooperation between China and the regions along the MSRI, mainly containing Asian, African and European countries. The blue Silk Road will be the focus not only of the Chinese government but also of the other governments involved in the BRI. All these official plans offer a practical guide for establishing and improving cooperation between China and its partners.

3.5.3 Practical projects

Central to the BRI is infrastructure development enabling and supporting the new connectivity between China and other participants. Under top-level direction and in the win–win cooperation spirit, the BRI has materialized in real and practical project initiatives. For example, from 2015 to November 2017, the Bank of China implemented credit support, amounting to approximately US$98 billion, to various BRI countries. Also, some major projects for the new Maritime Silk Road have been initiated, such as the building of enormous infrastructure, along with establishing some free trade agreements and bilateral investment treaties. For instance, China will make available more than 20,000 so-called connectivity professionals along the MSRI in the coming five years [36]. Another important activity is the improvement of standardization in building, proposed in many official documents and agreements. The representative document is the 'Action Plan on Connecting the Belt and Road by Standards (2018–2020)' issued in 2018. For this purpose, China will deploy no less than 1000 plans or tasks for translating Chinese standards to foreign versions and will strive to harmonize 2000 of the key technical indicators in the major domains [37]. These actions and implementations indicate the new Maritime Silk Road is an economic scheme rather than a political one, although some analysts cannot agree with the economic narratives [38]. As expected, the development and construction of the New Silk Road will be the main consideration and task not only of the Chinese government but also of the other BRI countries.

3.6 The effect on other countries of the New Silk Road

With advancement of the BRI, sustainable economic and trade cooperation between China and other countries has obviously increased, which has driven the New Silk Road to greater heights. Furthermore, the effects of the New Silk Road on the indigenous development of involved countries are noticeable and far-reaching. Until September 2016, China brought about strategic connectivity with 70 countries, regions and international organizations, achieving joint statements, bilateral and multilateral agreements and cooperation outlines as well as Memorandums of Understanding (MOU) [36]. Simultaneously, Chinese corporations established 56 overseas economic trade and cooperation zones in New Silk Road countries, with a total investment of US$18.55 billion and a total output value of US$50.96 billion, while contributing to the hosting countries taxes and fees of US$1.07 billion and providing 177,000 jobs [39]. There are different categories of countries involved, and the impact of the plans can be assessed as follows.

3.6.1 Effects on Africa

Originally, there were just four African countries (Egypt, Kenya, Ethiopia and Djibouti) [40] informally involved in the OBOR framework. However, until now, more African countries have expressed their wish to connect with China along the New Silk Road, in line with its '2063 Vision' of Africa [41]. In 2015, China and a number of African governments signed the MOU regarding constructions in the framework of the 'Three Networks & One Industrialization' (high-speed train network, expressway network and regional airline network as well as industrialization) to promote African infrastructure and equipment. At the end of the same year, Xi was present at the China and Africa Forum and proposed the 2016–2018 China and Africa cooperation plan, indicating support of US$60 billion for its implementation.

Egypt is an indispensable participant in the BRI, as the Suez Canal connects the Indian Ocean with the Mediterranean Sea [42]. In 2016, a top-level official MOU between China and Egypt was signed regarding the New Silk Road, the first formal agreement under the BRI in Africa [42]. For the benefit of the New Silk Road, the Central Business District (CBD) project for a new capital of Egypt was started, involving some US$45 billion, with the new residence planned to be opened in 2018 [42, 43]. The project itself has provided thousands of jobs in Egypt, while the new capital will give the

Egyptians greater opportunities to share in the benefits of new technology, talent development, capital accumulation and energy provision.

China invested US$4 billion in building the first international electric railway between Djibouti and Addis Ababa. At the same time, the railway linking the capital of Nairobi with the port city of Mombasa in Kenya costing US$3.8 billion has been constructed and began operation [40] in 2017. The flows of goods and trade involved in the port business can directly benefit from this line, while indirectly creating more job opportunities as well as further raising the income of the people.

Chinese investments in Tanzania have risen to US$6.6 billion by 2017 [44], with an annual US$4.6 billion increase in trade between the two countries [45], as announced by Lin Zhiyong (2017), Chief Commercial Representative at the Chinese Embassy. The expansion of the Port of Bagamoyo in Tanzania on Africa's eastern coast and the creation of a connected SEZ is an example [46]. Until now, China, as the largest investor and biggest business cooperator in Tanzania, has contributed to Tanzania's modernization and economic growth, although distributional effects require attention [46].

With the BRI, we have reason to believe that African countries have gained more resources for environmental upgrading, governance adjustment and industrial development [40]. Consequently, cooperation between China and Africa plus the projects under the OBOR initiative will be far-reaching and proof of the Chinese win–win idea.

3.6.2 Effects on ASEAN countries

Among the BRI countries, the number of Asian countries accounts for almost 50% of the total, so they are central to Chinese foreign policy and strategies to strengthen the relationships between China and other Asian countries. The ASEAN is an important regional organization in Asia, consisting of Brunei Darussalam, Cambodia, Indonesia, Lao PDR, Malaysia, Myanmar, Philippines, Singapore, Thailand and Vietnam. When the ASEAN–China free trade agreement was established in 2001 [47], the China-ASEAN (10 + 1) new relationship caught the attention of the world. From then on, the ASEAN plus China has become a significant cooperation framework in the regional economy [48].

The 10 + 1 relationship is a peaceful and stable environment for regional prosperity promotion based on political trust, according to Li Keqiang's statement at the 18th China–ASEAN Summit in 2015 [48]. The 10 + 1 members have complete independence in their national decision-making, confirming their actual equality. Economic cooperation is the core and purpose of the China–ASEAN relationship.

Within the 10 + 1 collaboration along the New Silk Road, foreign direct investment (FDI) can be seen as key. According to the 2016 China's Outward Investment Report [49], in 2015, the flow and stock of FDI of the BRI countries accounted for US$18.93 billion and US$115.68 billion, respectively. Within the FDI stock, investment in the ASEAN countries in 2015 was about US$47.11 billion, with 40.8% [50] of the total stock investment concentrated on business services, manufacturing, energy and financial companies.

In order to enlarge cooperation and communication on behalf of the multilateral economy and trade under the 10 + 1 friendship, many ASEAN countries stimulate higher-value industry development [51] to underpin their free trade strategies. Until now, the product trade structure has not been restricted to raw materials and chemicals between China and the ASEAN countries, higher technological and more valuable goods are also being transported among the countries, including smartphones, computer components, machinery and so on [52]. China has made a significant close connection with the ASEAN countries and is planning to enhance the trade agreement with the ASEAN for the 2020–2025 period [26].

AIIB investment projects in the ASEAN countries covered all domains of their domestic development, including slum-upgrading, flood control, natural gas infrastructure, broadband networks, electric systems and so on [53]. More strikingly, half of the core projects of international cooperation under the BRI are related to the ASEAN countries, especially Malaysia, Lao PDR and Myanmar. These projects mostly concern transportation, infrastructure and industry, which are aimed at improving trade as well as living standards.

The representative successful case is the construction and development of the Cambodian Sihanoukville Special Economic Zone (SSEZ). The SSEZ has been designed and constructed by the Chinese and Cambodian governments to achieve win–win cooperation. There are 118 companies in the economic zone with employment for 20,000 people until the end of 2017. The companies and people in Sihanoukville City benefit from SEZ projects and the cooperation between the two governments, obtaining 'five passings' plus 'one flat' (road passing, electricity passing, water passing, communication passing and pollution discharge, plus the ground flat) [54]. Under the New Silk Road, the SSEZ will be developed further, while serving as the model for bilateral cooperation and 10 + 1 relationship.

In summary, along with the more advanced economies in the ASEAN and other Asian countries [51] and China's economic growth, Chinese relations and links with Southeast Asian nations are moving into a new direction in regional economic development [47, 48].

3.6.3 Effects on Europe

As a pivotal strategic partnership of cooperation, the China–Europe relationship plays a crucial role in China's foreign policy. Within the BRI strategy and blueprint, European countries are the end-point and the largest trade partners for the long-term development of China and Europe. Since the OBOR initiative has been proposed, most European countries are open to it and are active in responding to the strategy and participating in many projects. Beginning with the 'Joint Document' resulting from the China-Central and Eastern European Countries (CEEC) Ministerial Meeting on Promoting Trade and Economic Cooperation Meeting in 2014 [55], the CEEC provided the first formal response to the OBOR programme. In the same year, Greece also expressed its support and participation in the MSRI during the successive visits of the Chinese Prime Minister Li Keqiang and President Xi Jinping. The 2015 MOU between China and Hungary followed, being the first BRI connective document [56] signed to support the Chinese initiative and integrate resource advantages. Further, during the 2017 BRF meeting, the MOU between the United Nations Economic Commission for Europe (UNECE) and the National Development and Reform Commission of China (NDRCC) was signed for the purpose of intensified public–private partnership (PPP) cooperation. The top-level link under the BRI is beneficial for both China and Europe. Against the background of weak world economic recovery and lagging global trade, the development trend in China–Europe trade is promising, with a 15.5% increased total trade value during the first four months of 2017 compared with the same period in 2016 [57].

Above all, the cooperation between China and the CEEC plays a major role in the New Silk Road programme. The China–CEEC cooperation was first established in Warsaw, Poland, in 2012, well known as the '16 + 1' cooperation framework. Under the OBOR strategy, the '16 + 1' framework has resulted in accelerated development in trade, investments, transportation and infrastructure construction during the past five years [58]. Taking Poland as an example, it is China's biggest trade partner in the CEEC region and the first member country of the AIIB – it is called the 'Showcase of the CEEC region' by Xi Jinping. The bilateral trade value between China and Poland increased from US$14.8 billion in 2013 to US$17.6 billion in 2016 [57]. After Xi's visit to Poland in 2016, and in line with the BRI, the two countries expressed belief and hope in cooperation in the domains of industry, agriculture, renewable energy and innovation [59]. For instance, the China Ever Bright International Limited enterprise achieved the largest Chinese acquisition in the environmental industry in the CEEC through the €123 million acquisition of Poland's leading waste treatment company NOVAGO

[60]. Large-scale projects have significantly improved job opportunities and increased fiscal revenues for the CEEC countries. Even more, markets and trade benefits have been enhanced, and infrastructure has been updated under OBOR international cooperation [56].

In addition, Greece is another main focus country with the New Silk Road, as it is situated in southern Europe and the Mediterranean. The cooperation and construction in the port of Piraeus have been one of the major MSRI projects so far [55]. Greece is considered as a significant gateway linking China to Europe, and Piraeus was called 'the pearl port' of the Mediterranean Sea by Li Keqiang on his visit to Greece [61]. The Chinese government made a plan for Piraeus as part of the new Maritime Silk Road, formally written into official top-level plans in 2015 and 2017. Consequently, Chinese funding, banks and companies have given large investments in infrastructure, logistics and shipping involving China's Export–Import Bank (CEIB), SRF, COSCO, Huawei and so on. The series of construction and investments help to promote the long-term recovery of Greece's weak trade and economy, as well as contributing to the land–sea express routes in Eurasia [61]. Also, Greek prosperity will benefit, while to some extent, the refugee crisis can be addressed by providing employment opportunities [62].

3.7 Discussion and conclusion

We have seen that China's recent opening-up actions are just a new stage in a long historical process showing a back and forth movement when it comes to closing and opening its borders for foreign trade and exchange in general. It is not difficult to conclude that greater openness in industries, regions and provinces will bring more success and competitiveness. Not only on behalf of the regional development in the Chinese SEZs but also to support the new global strategy and the Chinese opening-up foreign policy will be maintained and strengthened, providing more opportunity for prosperity in China. Obviously, the BRI aims to promote bilateral and multilateral partnerships in Asia in order to consolidate China's leadership status in the region and strengthen China's continued efforts in the field of national rejuvenation as a comprehensive national power. Meanwhile, the number of projects along the New Silk Road has been increasing, simultaneously improving the cooperation between China and the involved countries. We have already discussed some indigenous effects in other countries and will continue our discussion about the benefits and drawbacks below.

Above all, the BRI is a cooperation framework with many other countries rather than with individual ones, focusing on shared economic revitalization and growth. Actually, the Chinese government has formulated and signed the outlines of several economic corridors to be established between China and other countries. From the BRI point of view, these plans can be understood as a format for new spheres of trade [38, 63], giving rise to new regional governance. There is no doubt that the New Silk Road will give regional economies and the global economy as a whole more chances and choices.

Furthermore, the New Silk Road is a comprehensive win–win framework involving infrastructure, facilities, trade and investment, production capacity, financial cooperation and ecological and environmental protection, as well as exchanges in cultural, social and other fields [35]. As such, it has attracted increasing interest from government officials, academia and industry. By the end of 2016, Chinese enterprises had invested US$18.5 billion in 56 economic and trade cooperation zones in 20 BRI countries [35]. Also, the maritime industry contained many shareholding contracts for port investment along the Maritime Silk Road route. Through official documents and practical business experiences, two economic strategies are clear, one relating to economic prosperity at home, the other taking responsibility and leadership for regional and global economic recovery and growth.

However, the BRI is a long-term and international initiative since it is not only China's own project but also requires many other countries to cooperate and participate. Therefore, some questions, doubts, challenges and risks can be expected and should not be overlooked.

First, some countries are afraid of China's 'soft power.' This has to do with geopolitical aspects and perceptions of China's foreign strategy. But under official documents and agreements (MOUs), China respects the independence of other partners and retains its peaceful attitude. 'Soft power' refers to economic, peaceful and responsible development and cooperation rather than geopolitical domination or even military occupation.

Second, raising and collecting enough funding for the necessary investments is a big problem for all countries involved in the BRI. That applies to the Chinese government and businesses as well. At present, although many projects have received investment sums from the AIIB, SRF and other international or regional banking and funding organizations, the financing gap is still enormous, especially for the developing and less developed countries along the New Silk Road [64]. In the future, better funding methods are supposed to be taken into consideration for BRI projects.

Furthermore, because the New Silk Road is a long-term plan with many countries participating, how to balance the relationships among the members is a great challenge for China's continued opening-up policy. It can be expected that, in the future, more plans, agreements and cooperation will follow in the foreign policy of Xi Jinping and his leadership.

To conclude, new open foreign policies result from a combination of self-interest and moral calling, of domestic economic reasons and foreign political aspirations. The Belt and Road Initiative is a far-reaching strategy with decades of regional and worldwide implications. Nowadays, with its contributions to international peace, such as the peace fund established for peacekeeping, military deployment for regional disputes and being a primary provider of global security [21], China shows confidence in its further domestic development and new foreign policies. Great efforts are necessary to seize all promising opportunities, while the impacts, both positive and negative, should always be carefully mapped, mitigated if necessary and distributed!

Acknowledgements

Research for this paper was supported by the China Scholarship Council (CSC NO. 201707167030). It was part of the 'The Dao of Responsible Port Innovation' project at Delft University of Technology, the Humanities and Social Sciences Project of Harbin Normal University (SXP2014-07) and the Heilongjiang Philosophy and Social Sciences Plan Project (16JYC05).

References

[1] Full text of Xi Jinping's report at 19th CPC National Congress, Xinhua. http://www.xinhuanet.com/english/special/2017-11/03/c_136725942.htm. Accessed on: 3 November 2017.

[2] Huang, Y., Understanding China's Belt & Road initiative: Motivation, framework and assessment. *China Economic Review*, *40*, pp. 314–321, 2016.

[3] Cogman, D., Gao, P., Leung, N., Making sense of Chinese outbound M&A. https://www.mckinsey.com/business-functions/strategy-and-corporate-finance/our-insights/making-sense-of-chinese-outbound-m-and-a. Accessed on: 21 May 2021.

[4] Vision and Actions on Jointly Building Silk Road Economic Belt and 21st-Century Maritime Silk Road. Issued by the National Development and Reform Commission, Ministry of Foreign Affairs, and Ministry of Commerce of the People's Republic of China, with State Council authorization. http://www.chinese-embassy.org.uk/eng/zywl/t1251719.htm. Accessed on: 2 July 2021.

[5] Vision for Maritime Cooperation under the Belt and Road Initiative. Xinhuanet. http://www.xinhuanet.com/english/2017-06/20/c_136380414.htm. Accessed on: 20 June 2017.

[6] Doshi, R., Xi Jinping just made it clear where China's foreign policy is headed. https://www.washingtonpost.com/news/monkey-cage/wp/2017/10/25/xi-jinping-just-made-it-clear-where-chinas-foreign-policy-is-headed/?noredirect=on&utm_term=.e8b8d7b674d7. Accessed on: 25 October 2017.

[7] Profile: Xi Jinping and his era, Xinhua. http://www.xinhuanet.com/english/2017-11/17/c_136758372.htm. Accessed on: 17 November 2017.

[8] Work Together to Build a Community of Shared Future for Mankind. Xinhuanet. http://www.xinhuanet.com/english/2017-01/19/c_135994707.htm. Accessed on: 2 July 2021.

[9] Lin, F., *Opening Up and Chinese National Rejuvenation*. PhD thesis, Hunan Normal University, p. 46, 2009. (In Chinese.)

[10] Zhong, K., Historical reconsideration on China opening up in modern times. *Journal of Guangdong Education Institute*, 26(6), p. 2, 2006. (In Chinese.)

[11] Chen, Z., Nationalism, internationalism and Chinese foreign policy. *Journal of Contemporary China*, 14(42), pp. 35–53, 2005.

[12] Amako, S., China's diplomatic philosophy and view of the international order in the 21st Century. *Journal of Contemporary East Asia Studies*, 3(2), pp. 3–33. 2014.

[13] Cao, P., Opening Up and Chinese Development. PhD thesis, the Party School of the Central Committee of the CPC, p. 36, 1998. (In Chinese.)

[14] Loke, B., Between interest and responsibility: Assessing China's foreign policy and burgeoning global role. *Asian Security*, 5(3), pp. 195–215, 2009.

[15] Chang-Liao, N.C., China's new foreign policy under Xi Jinping. *Asian Security*, 12(2), pp. 82–91, 2016.

[16] Berkofsky, A., 'The Chinese Dream' and Chinese foreign and security policies: Rosy rhetoric versus harsh realities. *Asia-Pacific Review*, 23(2), pp. 109–128, 2016.

[17] Huang, H. & Zhou, C., China's strategic blueprint for its opening-up in the new era. *International Economic Review*, (4), pp. 23–49, 2013. (In Chinese.)

[18] Expo 2010, Wikipedia. https://en.wikipedia.org/wiki/Expo_2010. Accessed on: 10 November 2018.

[19] Medeiros, E.S., *China's International Behavior: Activism, Opportunism, and Diversification* (Vol. 850). Pittsburgh, PA: Rand Corporation, pp. 7–18, 2009.

[20] Ferdinand, P., Westward ho: The China dream and 'one belt, one road': Chinese foreign policy under Xi Jinping. *International Affairs*, 92(4), pp. 941–957, 2016.

[21] Poh, A. & Li, M., A China in transition: The rhetoric and substance of Chinese foreign policy under Xi Jinping. *Asian Security*, 13(2), pp. 84–97, 2017.

[22] Xinhua Insight: Xi's world vision: A community of common destiny, a shared home for humanity, Xinhua. http://www.xinhuanet.com/english/2017-01/15/c_135983586.htm. Accessed on: 15 January 2017.

[23] Li, L., China's manufacturing locus in 2025: With a comparison of 'Made-in-China 2025' and 'Industry 4.0'. *Technological Forecasting and Social Change*, 135, pp. 66–74, 2018.

[24] The Input Capital and Value for Governing the Smog Observing through Big Data in Beijing, Sohu. http://www.sohu.com/a/122459577_398736. Accessed on: 24 December 2016. (In Chinese.)

[25] Spotlight: Xi's Switzerland tour shows Chinese wisdom, confidence in face of sluggish global economy, anti-globalization sentiment, Xinhua. http://en.people.cn/n3/2017/0120/c90000-9169338.html#.WIME1vMcVHo.twitter. Accessed on: 20 January 2017.

[26] Holslag, J., Unequal partnerships and open doors: Probing China's economic ambitions in Asia. *Third World Quarterly*, 36(11), pp. 2112–2129, 2015.

[27] Kaczmarski, M., Two ways of influence-building: The Eurasian economic union and the one belt, one road initiative. *Europe-Asia Studies*, 69(7), pp. 1027–1046, 2017.

[28] Sidaway, J.D. & Woon, C.Y., Chinese narratives on 'One Belt, One Road' (一带一路) in geopolitical and imperial contexts. *The Professional Geographer*, 69(4), pp. 591–603, 2017.

[29] A comprehensive review of the progress of the 'Belt and Road' (Volume 1 of the 'Belt and Road' Research Manual). https://mbd.baidu.com/newspage/data/landingpage?context=%7B%22nid%22%3A%22news_6776530133898112223%22%2C%22sourceFrom%22%3A%22wiseindex%22%7D. Accessed on: 1 July 2021. (In Chinese.)

[30] 17 Projects signed Silk Road Fund, involving the total investment of US$80 billion, China News Online. https://www.yidaiyilu.gov.cn/xwzx/gnxw/38973.htm. Accessed on: 8 December 2017. (In Chinese.)

[31] Yu, H., Motivation behind China's 'One Belt, One Road' initiatives and establishment of the Asian Infrastructure Investment Bank. *Journal of Contemporary China*, 26(105), pp. 353–368, 2017.

[32] Introduction to Asian Infrastructure Investment Bank (AIIB). https://www.aiib.org/en/about-aiib/index.html. Accessed on: 2 July 2021.

[33] Asian Infrastructure Investment Bank Articles of Agreement. AIIB. https://www.aiib.org/en/about-aiib/basic-documents/_download/articles-of-agreement/basic_document_english-bank_articles_of_agreement.pdf. Accessed on: 21 May 2021.

[34] Action Plan on Belt and Road Standard Connectivity (2015–17). Office of the Leading Group for the Belt and Road Initiative. https://www.yidaiyilu.gov.cn/wcm.files/upload/CMSydylgw/201702/201702150616033.pdf. Accessed on: 21 May 2021. (In Chinese.)

[35] Building the Belt and Road: Concept, Practice and China's Contribution. Office of the Leading Group for the Belt and Road Initiative. https://eng.yidaiyilu.gov.cn/wcm.files/upload/CMSydylyw/201705/201705110537027.pdf. Accessed on: 21 May 2021.

[36] Blanchard, J.M.F. & Flint, C., The geopolitics of China's Maritime Silk Road Initiative. *Geopolitics*, 22(2), pp. 223–245, 2017.

[37] Action Plan on Belt and Road Standard Connectivity (2018–2020). Office of the Leading Group for the Belt and Road Initiative. https://www.yidaiyilu.gov.cn/zchj/qwfb/43480.htm. Accessed on: 11 January 2018. (In Chinese.)

[38] Blanchard, J.M.F., Probing China's Twenty-First-Century Maritime Silk Road Initiative (MSRI): An examination of MSRI narratives. *Geopolitics*, 22(2), pp. 246–68, 2017.

[39] Spotlight: Belt and Road Initiative delivers benefits beyond its borders, says Chinese ambassador, Xinhua. http://www.xinhuanet.com/english/2018-05/24/c_137203842.htm. Accessed on: 24 May 2018. (In Chinese.)

[40] African View on the 'Belt and Road Initiative,' Caixin website. http://opinion.caixin.com/2017-06-09/101099945.html. Accessed on: 9 June 2017. (In Chinese.)

[41] New Opportunity for the Cooperation between China and Africa, regarding the 'Belt and Road Initiative' and '2063 Vision,' People's Daily Online. http://world.people.com.cn/n/2015/1204/c1002-27888645.html. Accessed on: 4 December 2015.

[42] ZiroMwatela, R. & Zhao, C., Africa in China's One Belt, One Road Initiative: A critical analysis. *ISOR Journal of Humanities and Social Sciences*, 21(2), pp. 21–34, 2016.

[43] The 'Belt and Road Initiative' has taken a root in Africa, providing the residence housing and cars, Maritime Silk Road and Business Newspaper Online. http://fjrb.fjsen.com/nasb/html/2018-01/25/content_1083716.htm?div=2. Accessed on: 25 January 2018. (In Chinese.)

[44] China becomes Tanzania's biggest trading partner, Xinhua. http://en.silkroad.news.cn/2017/0823/59844.shtml. Accessed on: 23 August 2017.

[45] China encourages further investment in Tanzania, China Daily. http://africa.chinadaily.com.cn/weekly/2017-08/25/content_31090832.htm. Accessed on: 25 August 2017.

[46] Van Giezen et al., this volume, Chapter 7.

[47] Chirathivat, S., ASEAN–China Free Trade Area: Background, implications and future development. *Journal of Asian Economics*, 13(5), pp. 671–686, 2002.

[48] Ploberger, C., One Belt, One Road: China's new grand strategy. *Journal of Chinese Economic and Business Studies*, 15(3), pp. 289–305, 2017.

[49] Report on Development of China's Outward Investment and Economic Cooperation 2016. Ministry of Commerce of the PRA. (In Chinese.)

[50] 'Belt and Road' linking to many countries, making the wide trade prospective, Sina. http://finance.sina.com.cn/roll/2017-06-05/doc-ifyfuvpm7493211.shtml. Accessed on: 5 June 2017. (In Chinese.)

[51] Morck, R. & Yeung, B., China in Asia. *China Economic Review*, 40, pp. 297–308, 2016.

[52] Len, C., China's 21st Century Maritime Silk Road initiative, energy security and SLOC access. *Maritime Affairs: Journal of the National Maritime Foundation of India*, 11(1), pp. 1–18, 2015.

[53] Asian Infrastructure Investment Bank, AIIB; Belt and Road Portal. https://www.yidaiyilu.gov.cn/zchj/rcjd/958.htm. Accessed on: 28 September 2016.

[54] The economic contribution rate of the SSEZ to the local economy was exceeded 50%, absorbing more than 100 companies to settle in and providing employment for more than 20 thousand, Belt and Road Portal. https://www.yidaiyilu.gov.cn/xwzx/roll/43815.htm. Accessed on: 12 January 2018.

[55] Zhang, J. & Chen, Z., China–EU Cooperation under the 'Belt and Road Initiative': From a two-level EU perspective. *World Economics and Politics*, 11, pp. 36–52, 2015.

[56] Minghao, Z., The Belt and Road Initiative and its implications for China-Europe relations. *The International Spectator*, 51(4), pp. 109–118, 2016.

[57] Opportunities were Provided by the 'Belt and Road Initiative' for European Rejuvenation, People's Daily Online. http://world.people.com.cn/n1/2017/0510/c1002-29265111.html. Accessed on: 10 May 2017. (In Chinese.)

[58] The Budapest Guidelines for Cooperation between China and Central and Eastern European Countries, Xinhua. http://www.xinhuanet.com/

world/2017-12/01/c_1122039253.htm. Accessed on: 1 December 2017. (In Chinese.)

[59] Renewable energy cooperation and investment and financing between China and Poland along the 'Belt and Road'. https://www.huanbao-world.com/foreign/107771.html. Accessed on: 2 July 2021. (In Chinese.)

[60] International Business. Everbright International. https://www.ebchinaintl.com/sc/business/international.php. Accessed on: 21 May 2021.

[61] Casarini, N., When all roads lead to Beijing: Assessing China's New Silk Road and its implications for Europe. *The International Spectator*, 51(4), pp. 95–108, 2016.

[62] Tzimitras, V., Towards a low emission port: Development of a decision making tool for technologies for the reduction of the shore-related emission footprint of existing ports in respect of stakeholders' values. Master's thesis, Delft University of Technology, 2018.

[63] Clover, C. & Hornby, L., China's Great Game: Road to a new empire. *Financial Times*, 12, p. 1, 2015.

[64] Aoyama, R., 'One Belt, One Road': China's new global strategy. *Journal of Contemporary East Asia Studies*, 5(2), pp. 3–22, 2016.

[65] Belt and Road International Cooperation, Wikimedia Commons, https://commons.wikimedia.org/wiki/File:Before_the_beginning_of_the_Belt_and_Road_international_forum.jpg. Accessed on: 6 December 2018.

CHAPTER 4

Drawing lessons from eco-city to eco-port development in China: A policy perspective

Haiyan Lu[1], Martin de Jong[1,2,3] & Yan Jia[4]

[1]Faculty of Technology, Policy and Management, Delft University of Technology, The Netherlands
[2]Erasmus School of Law, Erasmus University Rotterdam, The Netherlands
[3]School of International Relations and Public Affairs, Fudan University, China
[4]School of Public Policy and Management, Tsinghua University, China

Abstract

Recently, the Sustainable Development concept has attracted more attention from Chinese cities. Ecological initiatives can be found in their infrastructure provision, and one specific result is eco- or green port development. Although eco- or green port development faces implementation barriers, most research focuses on ecological initiatives at the city level instead of on port development. As eco-port and eco-city development programmes share many implementation features in China, we explore which lessons can be drawn for eco-port development. We first review the literature on eco-city development. Then, we draw lessons from eco-city development bearing the status quo of eco-port development in China in mind. In the decision-making process, the support from higher levels of government and input from international actors can incorporate more sustainable initiatives in the design and planning of eco-port development. In the implementation process, the involvement of municipalities and citizens can improve the engagement and contribution at the local level. Finally, clear and useful evaluation indicators about the management process help develop the eco-port step by step.

Keywords: Eco-port, green port, eco-city, policy process, decision-making, implementation, lesson drawing

4.1 Introduction

After the opening-up and reform policy, China's port sector achieved leapfrog development, accompanied by a rapid increase in manufacturing and international trading. There were 362 ports with one million tonnes of capacity in 2016 [1]. In 2017, China owned six container ports in the world's top ten port list. Driven by the globalization process and the export growth in China, port expansion will continue into the future. However, the continuous construction and expansion of ports will inevitably have a negative impact on their surrounding environment.

With the further expansion of the scales of ports, the Sustainable Development concept has attracted more attention in port development, especially for the planning and construction process. Researchers began to study the Sustainable Development of ports and analysed demonstration cases in various countries [2, 3]. Facing environmental deterioration and the energy crisis, ecological ports, also known as green ports, are a model to consider for both environmental requirements and economic benefits [4]. Inspired by environmental sustainability related to aspects of growth limits and development scales [5], green ports also focus on the long-term strategy for port development, while considering the surrounding environment and climate change [6]. These targets demand that ports include environmental aspects into their operations [7].

Although relevant policies and regulations have been issued by governments governing the construction of eco-ports [8], implementation does not seem to be a target. In the planning and construction process of port development, developers and operators paid less attention to the environmental aspects than the economic targets. Their focus remains on improving the functional capability and market competitiveness of ports. Although relevant legislation on environmental protection has been introduced in port development, the monitoring mechanism of these environmental initiatives has not been established [9]. In China, operators are not fully aware of the integration of environmental protection in port management and operation [10]. Additionally, port authorities and operators provide limited financial and technical input for eco-port development [11]. Even in the pioneer ports pursuing ecological initiatives, sustainable concepts have remained a paper tiger. Unclear evaluation indicators and weak cooperation between the public and private sectors are the main barriers in the implementation process [12].

The difficulties in implementing sustainable concepts are not only limited to port management but can also be found at the city level, such as in eco-city development. The concept of the eco-city was initially promoted by urban planners, aiming for a balance between environmental impact and economic benefits on the city scale [13, 14]. Its main principles include revising land-use priorities to create compact, diverse, green, mixed-use communities near transit nodes and support local agriculture, urban greening projects and community gardening [15]. Later on, these ecological initiatives were promoted in urban planning at the community and municipal levels in different countries [16].

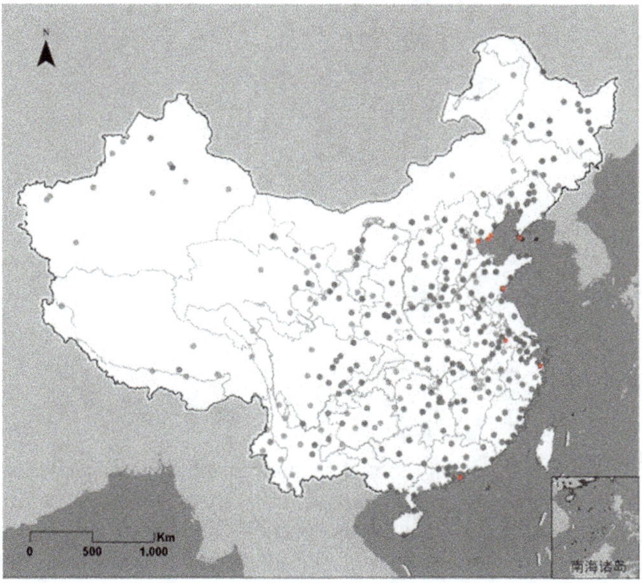

Figure 4.1: The low-carbon eco-city programmes in China [17] and the eco-port pilots in 2016.

Both eco-port and eco-city development in China are highly affected by the policies of the national government. The eco-city programme was launched earlier than the eco-port programme by the national government. In 2003, the Ministry of Environmental Protection (MEP) initiated an 'eco-cities' programme, and more than 100 cities claimed to be eco-cities in the 2010s (see Figure 4.1). After almost a decade, the targets for the ecological development of ports were proposed in the *Shipping Transport Low Carbon Development 12th Five Year Plan (2011)*. In 2015, the Ministry of Transport also issued the 'Port Energy Management Pilot' programme, which aimed at encouraging port enterprises to reduce their energy use systematically. In doing so, both policies assumed the role of city patrons and port enterprises across the country in becoming exemplars for Sustainable Development (see the red points in Figure 4.1). The concept of the ecological or green port has been adopted by most large port enterprises in China. One reason is that port operators are state-owned enterprises, which are easier to influence with government policies.

Operating at a larger scale, eco-city development and policies have been studied more than eco-ports. From 1996 to 2018, 127 articles, reviews or books in Scopus appeared on the topic of eco-city development, whereas only 33 studies discussed eco- or green ports (see Figure 4.2). This phenomenon cannot

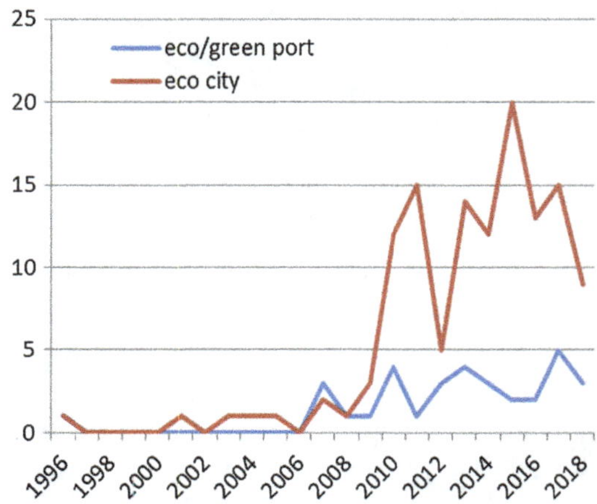

Figure 4.2: Articles, reviews or books about eco-cities and eco-/green ports in Scopus (the number of eco-city studies is based on the combination of keywords 'eco cit*' and 'polic*', and the other based on the keyword combination of 'eco/green port*' and 'polic*').

only be explained by the short policy implementation period of eco-ports, it is also due to the limited number of contributions from various disciplines. The scholars interested in eco-port development are mainly from technology and project management. In contrast, more scholars from interdisciplinary fields contributed to the debate in the policy process of eco-cities, such as urban planning, geography and public policy (see Figure 4.3). As both eco-port and eco-city development face some implementation barriers, the abundant number of studies about eco-cities from the perspective of policy process prompts us to *understand how lessons can be drawn for eco-port development.*

In this chapter, we will introduce the differences and similarities of the policy process between eco-cities and eco-ports. Then, we discuss the possibilities for lesson drawing from eco-city and eco-port development. Finally, conclusions are provided at the end.

4.2 Differences and similarities of the policy process between eco-cities and eco-ports

4.2.1 Differences

4.2.1.1 Scale and function

Eco-cities are in essence a type of newly built area incorporating technological and ecological features to create better environmental, social and economic

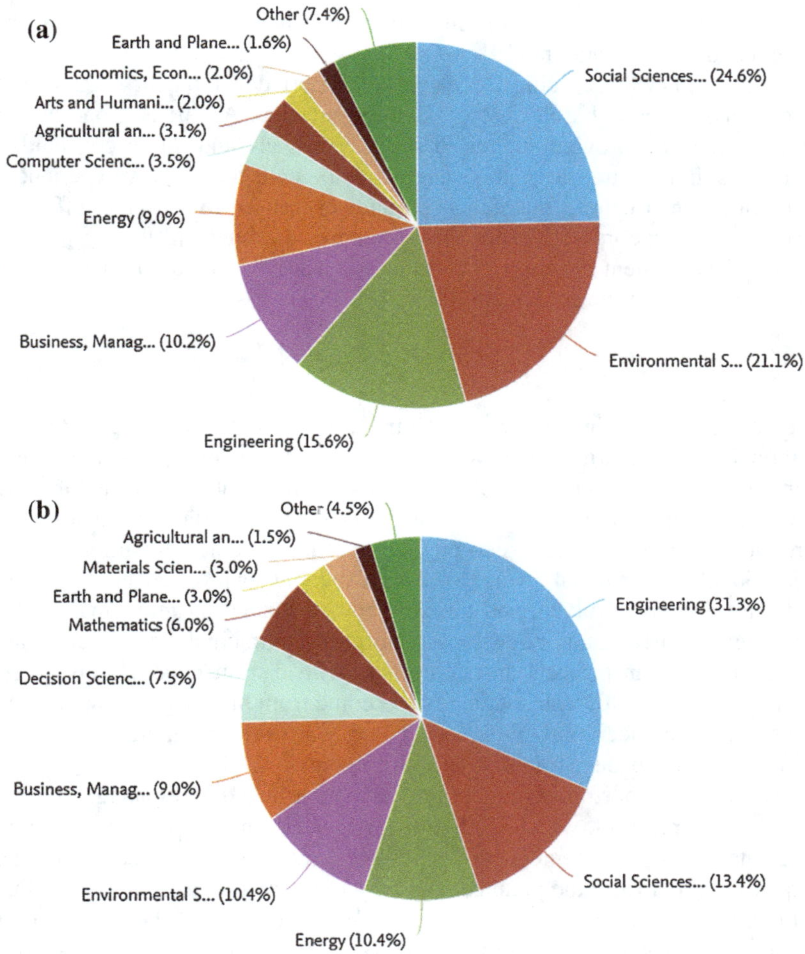

Figure 4.3: (a) Subject areas of studies related to 'eco-city' and 'policy'; (b) subject areas of studies related to 'eco-port' or 'green port' and 'policy.'

conditions. In China, eco-cities were initially focused on improving environmental quality and adopting green buildings and clean technologies in projects [18]. Against a background of clean technology development, cities in China more recently have begun to explore solving urban problems through applying technology solutions in infrastructures, such as buildings, the traffic system and waste treatment [19].

Compared with eco-cities, eco- or green port development occurs on a much smaller scale, based on the existing port area. A port is a harbour or an area that can provide shelter to ships or vessels and allow constant or periodic shipment transactions. Some scholars state that a port experiences planning, construction, maintenance, and operation to provide effective services while incorporating the surrounding environment [20]. The process of port development can impact air quality, cause water pollution and even interfere in the marine ecosystem [20–23].

4.2.1.2 Actors

The actors involved in eco-cities in China are different levels of government (national, provincial, municipal and district), developers and financiers (either as public or private enterprises), architecture and consultancy firms hired by governments or developers and in some cases non-governmental actors and foreign governments. Some eco-city programmes are led by ministries at the national or provincial level. However, these higher-level governments are often not directly involved, and the projects are funded by municipal and/or district governments and/or developers. Normally, municipal governments play the leading role in projects and coordinate with district governments at the early stages [24]. Although spatial visions come from higher governments, the complexity and uncertainty of the local context for these urban projects require organizational capacity from district governments.

Before 1978, port management had been centralized and managed by the Ministry of Transport, which made important decisions for port development in China. This management strategy is regarded as less effective in the transition period from the planned economy to the market economy in China. Following the management practices in Western countries, the task of port management was allocated to the municipal government level [25]. However, in the beginning stage of this reform, the municipalities acted both as the regulator and the port manager in port development.

In the 1980s, the national government promoted a clear policy to allocate responsibilities between government and enterprises. Port enterprises have been provided with more rights in port development, especially regarding the ports' business performance. In the 1990s, foreign companies were also encouraged to invest and participate in the construction and operation of port facilities in China [26]. In recent years, port enterprises have also added a new vision to port development, such as developing technology and smart ports. These visions focus on how to make the best decisions in port services with the help of technology and innovations.

4.2.2 Similarities

In the literature, the policy process of the eco-city and eco-port is influenced by top-down policy initiatives in China. A similar policy process makes drawing lessons possible.

4.2.2.1 Top-down policy initiatives

In China, as environmental deterioration has become acute in recent years, many policies are issued from the national level to set targets or guidelines for local governments. Both 'eco-cities' and 'eco-ports' are promoted through government policies in a top-down manner. Since the implementation of national policies at the local level may face similar problems in the Chinese context, this top-down policy initiative provides a possible angle to understand eco-city and eco-policy processes in China. The financial and policy support from national governments play a crucial role in eco-city development in China [27]. In eco-port development, the encouragement to go beyond the minimum environmental standards also requires incentives and support from higher governments [28].

4.2.2.2 Styles in the policy process

In the literature on environmental policies, many scholars have analysed the policy process as happening in stages and studied how environmental initiatives are incorporated at each stage. Joss has classified eco-city projects according to their development phase, including 'at the planning stage,' 'under construction' or 'implemented' [29]. Yin further investigated how environmental concerns or objectives are integrated and how they become part of the fundamental premises in each stage of eco-city project development, such as decision-making and implementation [24]. In the Chinese political context, the policy stages are adopted by governments in policy design. For instance, the Provincial Port Regulation in China is designed according to different policy stages, including planning, construction, maintenance and operation. It is possible to follow this trend to classify the dilemmas of eco-port development by the different policy stages they belong to [12]. Above all, although the policy stage model is criticized for oversimplifying the policy process [30], it offers useful clues in describing policymaking practices in both eco-city and eco-port development in China.

4.2.2.3 The possibilities for lesson drawing

The previously mentioned similarities offer the opportunity to draw lessons between eco-city and eco-port development. Due to larger scales, the eco-city covers more sections, such as transportation, housing and industrial development. The various issues in these sectors involve more stakeholders than in eco-port development. In addition, eco-cities require more extended planning and implementation periods, varying from 5 to 30 years, which is generally longer than for eco-ports. As the eco-city development in China involves larger scales, more involved actors and longer experimental periods, the experience in dealing with complexity must be substantial.

4.3 The review of the eco-city policy process

This section will review the studies on eco-city development in China, especially from the perspective of public policy. In the 1980s, with increasing environmental pressure amidst fast economic growth, the development of the eco-city became prominent with support from the national government. In the 2000s, more scholars investigated eco-city policies and practices in single or multiple case studies, such as the projects in Tianjin, Suzhou, Tangshan and Shenzhen. Inspired by the classification of eco-cities as a policy process [29], we summarized the previous research on eco-cities in China from a public policy perspective, including decision-making and implementation.

4.3.1 The decision-making process

In China, governments have faced enormous international pressure to tackle environmental deterioration. In the past few decades, the national government has explored the approaches for eco-city development by providing municipal governments with the autonomy to carry out experiments with supporting policies. As for municipal governments, the approval of a flagship or demonstration project also helps them to restructure their economies, such as more investment in high-tech industries and the real estate sector [31]. Regarding eco-cities as policy experimentation in dealing with environmental challenges, some scholars have investigated the main factors that influence policy outcomes of these eco-city projects [31–34]. The support of higher government and international policy transfers are the factors mentioned most by scholars in eco-city studies [26, 35].

Miao and Lang (2014) investigated the Dongtan and Tianjin eco-city projects and found that project developers respected China's unitary and hierarchical organization when promoting eco-city initiatives, although under different regulatory structures. They argued that the support from higher-level governments played a positive role in promoting eco-city development. On the other hand, some scholars also doubted the top-down policy approach, which may obscure required participation at the local level. After studying the low-carbon action plans adopted in low-carbon cities, Khanna et al. concluded that comprehensive guidance from the national level limited input from stakeholders and the public, which made it difficult to include the various aspects of urban development in the planning [36]. De Jong et al. also found that citizens and citizen groups have little influence on decision-making processes regarding eco-city projects in China in comparison with practices in Europe [35].

Meanwhile, scholars also stated that eco-city development in China has coped with various foreign and domestic policy ideas in a modular way. Some of them are positive about the role of external political and economic actors in decision-making on eco-city development. For instance, Miao and Lang stated that an influential transnational investor could secure crucial support from the national government in eco-city projects [37]. Additionally, the developers in some eco-city projects, which are joint ventures between Chinese international companies, have brought financial and technical input into the projects [38]. Moreover, international companies also provide imaginative and innovative designs to decision-makers in eco-city development [31]. However, some scholars doubt the effectiveness of these quick-fix green urban solutions around the world, especially in handling contradictions and political struggles in the specific Chinese context [39]. Above all, eco-city development requires a cautious and selective approach that reflects a more generic Chinese tradition of institutional bricolage [40], and most scholars agree that these learning experiences imply important lessons.

4.3.2 The implementation process

As most of the environmental policies are initiated by national governments, some scholars explain the implementation gap by insufficient monitoring. For example, the low-carbon economy policy is issued by the national government, and eco- and low-carbon cities are critical pilots on the local level. Some scholars have compared the targets, scopes and supporting measures of these national pilots, and they concluded that the key barriers include implicit definitions of what the low-carbon city is, complexity and confusion of having several parallel programmes and insufficient supporting policies and market-based instruments [36, 40, 41]. Other scholars also state that unclear visions or targets and ineffective indicators are the main barriers to the promotion of sustainable initiatives [42].

Still, other scholars examined the behaviour of local government and attributed implementation problems to their strong economic ambitions and limited environmental interest [42–44]. Some scholars believe in the sincerity of Chinese governments to pursue ecological initiatives, but they are disappointed by the priority given to GDP growth by the municipal governments [40]. Municipalities collaborate with developers to gain land revenues in eco-city projects, which can be seen as the existence of stake-business urban growth coalitions as in the United States [39]. Some scholars even argue that the nature of eco-cities in China is just entrepreneurial land development [45]. Caprotti highlighted the land market, reclamation, dispossession and 'green grabbing' in eco-city development processes [33]. Wu expressed concern about the unexpected speed of eco-city development. Some projects are funded by bank loans, betting on high future land values from real estate development [31]. Above all, the entrepreneurial behaviour of municipal governments and excessive budget dependence on land concessions are the main barriers recognized by scholars.

4.4 Lessons drawn from eco-port development

Based on the review of the policy process for eco-city projects in China, we will attempt to draw lessons from eco-port developments. This section proposes the possible lessons to be drawn from the policy process regarding eco-cities.

4.4.1 Decision-making on eco-ports

Lesson one. The support from higher governments has a positive effect on the green concept in port development, especially when municipalities have limited financial capabilities.

After 1984, the centralized management system for ports was gradually adjusted to better incorporate local managerial systems. Many national ports were renamed as provincial ports, and the port authorities were relocated to municipal governments. State-owned enterprises were also separated from government bureaus to administrate and operate the ports. In recent years, port development has received less attention from the national government, compared with eco-city projects. National or provincial governments regard port development as a municipal-level issue. However, the financial capability of municipalities varies across China. As for second- and third-tier cities, the municipalities are still under pressure to pursue economic development and have fewer resources to support ecological initiatives.

As for the cities at a financial disadvantage, it is necessary to integrate port development into the policy package in specific city development programmes to gain support from the national government. For instance, infrastructure projects in Shenzhen International Low Carbon City have received subsidies from national and provincial Development and Reform Committees because they are a national-level low-carbon city pilot project. Subsidies or other policy support from the national government increases the willingness and capability of actors to strive for eco-targets, which can also be applied in eco-port development.

Lesson two. Proper international policy transfer has a positive effect on sustainability achievements in eco-port development.

In 1985, the State Council issued a policy to promote Sino–foreign joint ventures on harbour and port construction. The national government sensed the importance of introducing foreign capital, technology transfer and management knowledge in port development [46]. Following this, foreign companies have invested in the construction and operation of port facilities, as well as container terminals [25]. Above all, most of the international actors in Chinese port development are private sector players.

The international actors involved in eco-city developments are more complicated than in eco-ports, including different tiers of governments, non-governmental organizations, planners, research institutes and experts. On the international level, the World Bank and UN-Habitat have promoted these initiatives. On the national government level, the Sino–EU urbanization demonstration projects, Sino–US city development projects and Sino–Singapore cooperation have promoted many eco-city development projects. On the municipal government level, the C40 Cities Climate Leadership Group also involves some Chinese mayors to share best eco-city or low-carbon city practices [47]. Furthermore, non-governmental organizations, such as the World Alliance for Low Carbon Cities, actively promoted low-carbon city development.

Furthermore, in the high quality for eco-city projects in China, various types of planning competitions were organized to attract input from international planners and architects. Moreover, many scholars conducted research projects and provided their knowledge and insight on eco-city projects. Above all, different non-governmental organizations have participated in eco-city projects.

Compared with eco-cities, eco-ports enjoy less involvement from international organizations, national governments, non-governmental organizations and planners. The China Port Association has declared that green port development is one of its tasks, but limited cooperation or networking has been organized to communicate with international port authorities or enterprises. Additionally, there are few independent organizations to promote environmental initiatives. This has also been echoed in a previous finding that fewer scholars have been focused on research for eco-port development.

4.4.2 Implementation of eco-ports

Lesson three. Support from municipal governments can improve the implementation of eco-port development.

Port development is an integral part of city development. Visions for ports are easy to achieve when they are incorporated into urban development targets. For instance, Rotterdam Port observed that the municipal government emphasized it as an attractive place to live and work for citizens [47–49]. Rotterdam Port authorities endeavour to cooperate with the municipal government to seek common ground in adopting clean energy and reducing their ecological footprint. With the contribution from the port to the surrounding environment, the port also benefits from business opportunities and technological input from broader urban development. In this way, eco-ports can seek opportunities from the eco-city development process. The port companies can also benefit from knowledge input about Sustainable Development from designers, planners and experts in eco-city projects. Additionally, eco-port projects can improve citizen participation by co-organizing workshops or other interactive events with eco-city projects.

Lesson four. Clear evaluation indicators regarding the process can help achieve sustainability targets; the evaluation systems of the project can strengthen the motivation to reach environmental goals.

Clear evaluation indicators from a managerial point of view can divide eco or low-carbon targets into possible small steps. In eco-port development, ecological indicators have mainly focused on air pollution, liquid pollution, noise, solid waste and marine biology as well as aesthetic impacts [50]. The pitfall of these indicators is that less attention is paid to the management process. In contrast, some indicators in eco-cities have already incorporated the management perspective into their indicators [51]. In eco-port development, these lessons can be drawn by developing proper evaluation indicators for the management process.

Additionally, in China, some evaluation systems about eco-cities are issued by research institutes, such as the Chinese Academy of Social Science. These systems urge cities to be more concerned about their image and ecological rankings, such as being a liveable city and green city ranking lists. However, only a few ranking lists have been issued in the port sector, and most of the competition among ports still focuses on economic indicators. Specific programmes or ranking lists could be designed by some research institutes or other organizations to raise awareness of the ecological aspects of port development.

4.5 Conclusion

Eco-cities and eco-ports cover different scales, functions and involved actors. However, both are promoted by the national government through top-down policies. Their similar policy processes make it possible to draw lessons. Since eco-city development has a longer-term implementation period, this study focused on how eco-port development can learn from successful cases and also avoid pitfalls.

In the decision-making process, the support from higher government and input from international actors can strengthen more sustainable initiatives in the design and planning process of the eco-port development. In the implementation processes, the involvement of local stakeholders, such as municipalities or citizens, can improve engagement and contribution from the local level. Finally, transparent and effective evaluation indicators about the management process allow the eco-port to develop step by step. In a broader sense, evaluation systems from corresponding research institutes can put pressure on the ports to reach their ecological goal.

However, the previous discussion does not mean eco-cities in China are well developed without any pitfalls. Scholars also criticize eco-cities in China due to their unclear targets, limited market instruments and reduced citizen involvement. Eco-port development can also consider these implementation barriers and avoid them in the early stages of implementation.

Acknowledgements

The authors are grateful for the comments from Dr W. (Wim) Ravesteijn.

References

[1] Ministry of Transport. *Transport Development Bulletin*, 2016.

[2] González, M.M. & Trujillo, L., Efficiency measurement in the port industry: A survey of the empirical evidence. *J Transp Econ Policy*, 43(2), pp. 157–192, 2008.

[3] Wang, K., Ng, A.K.Y., Lam, J.S.L. & Fu, X., Cooperation or competition: Factors and conditions affecting regional port governance in South China. *Maritime Economics & Logistics*, 14(3), pp. 386–408, 2012.

[4] Lam, J. & Van De Voorde, E., Green port strategy for sustainable growth and development. In: *Transport Logistics for Sustainable Growth at a New Level: Proceedings of the International Forum on Shipping, Ports and Airports (IFSPA)*, pp. 417–427, 2012.

[5] Goodland, R., The concept of environmental sustainability. *Annu Rev Ecol Syst*, 26, pp. 1–24, 1995.

[6] Pavlic, B., Cepak, F., Sucic, B., Peckaj, M. & Kandus, B., Sustainable port infrastructure: Practical implementation of the green port concept. *Therm Sci*, 18(3), pp. 935–948, 2014.

[7] Becker, A., Inoue, S., Fischer, M. & Schwegler, B., Climate change impacts on international seaports: Knowledge, perceptions, and planning efforts among port administrators. *Clim Change*, 110(1–2), pp. 5–29, 2012.

[8] Chang, C.C. & Wang, C.M., Evaluating the effects of green port policy: Case study of Kaohsiung harbor in Taiwan. *Transp Res Part D Transp Environ*, 17(3), pp. 185–189, 2012.

[9] Bateman, S., Environmental issues with Australian ports. *Ocean and Coastal Management*, pp. 229–247, 1996.

[10] Xue, J., Li, Q., Wang, N. & Sun, G., Sustainable development in China's seaport: Trends and solutions. *J Liaoning Norm Univ Sci Ed.*, 2, pp. 228–231, 2008.

[11] Liu, C. & Wang, X. Problems and solutions in the development of green and low carbon in China's ports. *Shipp Manag*, 3, pp. 21–24, 2017.

[12] Chen, S., Wang, X., Guo, J., Feng, Z. & Yu, J., Problems and policy recommendations for China's green port construction. *China Transp Rev*, 7, pp. 19–21, 2016.

[13] Register, R., *Ecocity Berkeley: Building Cities for a Healthy Future.* Berkeley, CA: North Atlantic Books, 1987.

[14] Kenworthy, J.R., The eco-city: Ten key transport and planning dimensions for sustainable city development. *Environ Urban*, 18(1), pp. 67–85, 2006.

[15] Roseland, M., Dimensions of the eco-city. *Cities*, 14(4), pp. 197–202, 1997.

[16] Rapoport, E., Utopian visions and real estate dreams: The eco-city past, present and future. *Geogr Compass*, 8, pp. 137–149, 2014.

[17] Yu, L., Low carbon eco-city: New approach for Chinese urbanisation. *Habitat Int*, 44, pp. 102–110, 2014.

[18] Hult A. Swedish production of sustainable urban imaginaries in China. *J Urban Technol*, 20, pp. 77–94, 2013.

[19] Joss, S., Eco-cities: The mainstreaming of urban sustainability: Key characteristics and driving factors. *Int J Sustain Dev Plan*, 6(3), pp. 268–285, 2011.

[20] Tsinker, G.P., *Port Engineering: Planning, Construction, Maintenance, and Security*. Hoboken, NJ: Wiley, 2004.

[21] Liao, C.H., Tseng, P.H., Cullinane, K. & Lu, C.S., The impact of an emerging port on the carbon dioxide emissions of inland container transport: An empirical study of Taipei port. *Energy Policy*, 38(9), pp. 5251–5257, 2010.

[22] Ng, A.K.Y. & Song, S., The environmental impacts of pollutants generated by routine shipping operations on ports. *Ocean Coast Manag*, 53(5–6), pp. 301–311, 2010.

[23] Peris-Mora, E., Orejas, J.M.D., Subirats, A., Ibáñez, S. & Alvarez, P., Development of a system of indicators for sustainable port management. *Mar Pollut Bull*, 50(12), pp. 1649–1660, 2005.

[24] Yin, Y., Rader Olsson, A. & Håkansson, M., The role of local governance and environmental policy integration in Swedish and Chinese eco-city development. *J Clean Prod*, 134, pp. 78–86, 2016.

[25] Wang, J.J., Ng, A. & Olivier, D., Port governance in China: A review of policies in an era of internationalizing port management practices. *Transp Policy*, 11(3), pp. 237–250, 2004.

[26] Comtois, C. & Dong, J., Port competition in the Yangtze River Delta. *Asia Pac Viewp*, 48(3), pp. 299–311, 2007.

[27] de Jong, M., Yu, C., Joss, S., Wennersten, R., Yu, L., Zhang, X., et al., Eco city development in China: Addressing the policy implementation challenge. *J Clean Prod*, 134(Part A), pp. 31–41, 2016.

[28] Lam, J.S.L. & Notteboom, T., The greening of ports: A comparison of port management tools used by leading ports in Asia and Europe. *Transport Reviews*, 34, pp. 169–189, 2014.

[29] Joss, S., Eco-cities: A global survey 2009. *Ecology*, 129, pp. 239–250, 2010.

[30] Sabatier, P.A. & Christopher, W., Theories of the policy process. *Theories of the Policy Process*, pp. 189–220, 2007.

[31] Wu, F., China's eco-cities. *Geoforum*, 43, pp. 169–171, 2012.

[32] Yu, C., Dijkema, G.P.J., de Jong, M. & Shi, H., From an eco-industrial park towards an eco-city: A case study in Suzhou, China. *J Clean Prod*, 102, pp. 264–274, 2015.

[33] Caprotti, F., Critical research on eco-cities? A walk through the Sino-Singapore Tianjin Eco-City, China. *Cities*, 36, pp. 10–17, 2014.

[34] Joss, S. & Molella, A., The eco-city as urban technology: Perspectives on Caofeidian International Eco-City (China). *J Urban Technol*, 20(1), pp. 115–137, 2013.

[35] de Jong, M., Wang, D. & Yu, C., Exploring the relevance of the eco-city concept in China: The case of Shenzhen Sino-Dutch Low Carbon City. *J Urban Technol*, 20(1), pp. 95–113, 2013.

[36] Khanna, N., Fridley, D. & Hong, L., China's pilot low-carbon city initiative: A comparative assessment of national goals and local plans. *Sustain Cities Soc*, 12, pp. 110–121, 2014.

[37] Miao, B., Lang, G., A tale of two eco-cities: Experimentation under hierarchy in Shanghai and Tianjin. *Urban Policy Res*, 33(2), pp. 1–17, 2014.

[38] De Jong, M., Yu, C., Chen, X., Wang, D. & Weijnen, M., Developing robust organizational frameworks for Sino-foreign eco-cities: Comparing Sino-Dutch Shenzhen Low Carbon City with other initiatives. *J Clean Prod*, 57, pp. 209–220, 2013.

[39] Pow, C.P. & Neo, H., Modelling green urbanism in China. *Area*, 47(2), pp. 132–140, 2014.

[40] de Jong, M., China's art of institutional bricolage: Selectiveness and gradualism in the policy transfer style of a nation. *Policy Soc*, 32(2), pp. 89–101, 2013.

[41] Wang, C., Lin, J., Cai, W. & Zhang, Z., Policies and practices of low carbon city development in China. *Energy Environ*, 24(7–8), pp. 1347–1372, 2013.

[42] Liu, H., Zhou, G., Wennersten, R. & Frostell, B., Analysis of sustainable urban development approaches in China. *Habitat Int*, 41, pp. 24–32, 2014.

[43] Liu, L., Liu, C. & Sun, Z., A survey of China's low-carbon application practice: Opportunity goes with challenge. *Renew Sustain Energy Rev*, 15(6), pp. 2895–2903, 2011.

[44] Lo, K., China's low-carbon city initiatives: The implementation gap and the limits of the target responsibility system. *Habitat Int*, 42, pp. 236–244, 2014.

[45] Shiuh-Shen, C., Chinese eco-cities: A perspective of land-speculation-oriented local entrepreneurialism. *China Inf*, 27, pp. 173–196, 2013.

[46] Caprotti, F., Eco-urbanism and the eco-city, or, denying the right to the city? *Antipode*, 46(5), pp. 1285–1303, 2014.

[47] Li, Z., Galeano Galván, M.J., Ravesteijn, W. & Qi, Z., Towards low carbon based economic development: Shanghai as a C40 city. *Sci Total Environ*, 576, pp. 538–548, 2017.

[48] Port of Rotterdam. *Progress Report 2014 Port Vision 2030*. Rotterdam: Port of Rotterdam, 2014.

[49] Ravesteijn, W., Liu, Y. & Yan, P., Responsible innovation in port development: The Rotterdam Maasvlakte 2 and the Dalian Dayao bay extension projects. *Water Sci Technol*, 72(5), pp. 665–677, 2015.

[50] Lirn, T., Jim Wu, Y. & Chen, Y.J., Green performance criteria for sustainable ports in Asia. *Int J Phys Distrib Logist Manag*, 43(5/6), pp. 427–451, 2013.

[51] Joss, S., Tomozeiu, D. & Cowley, R., Eco-city indicators governance challenges.pdf. In: M. Pacetti (ed.) *The Sustainable City VII: Urban Regeneration and Sustainability*. Ashurst Lodge, UK: WIT Press, p. 1109, 2012.

Part II

Responsible Port Innovation: Practical examples

CHAPTER 5

Green development of the Port of Shanghai: Towards leadership for the Maritime Silk Road by setting a good example

Zhijie Li[1,2] & Wim Ravesteijn[2,3]

[1]School of Management, Harbin Institute of Technology, China
[2]Department of Values, Technology and Innovation, Delft University of Technology, The Netherlands
[3]Beijing Information Science and Technology University, China

Abstract

Increasingly, attention has been paid to the Chinese One Belt, One Road programme, not only when it comes to the economic impact, but also regarding the environmental impact, especially for the global climate. The Port of Shanghai, as a key connection on the Maritime Silk Road, has just completed its 'Three-Year Green Port Action Plan,' bringing it closer to its 'Green and Smart' port goals. Consequently, Shanghai could become a 'green' example for the ports along the OBOR. The main questions of this chapter are: what is the current development status in Shanghai and how can the port improve itself in such a way that it sets an example for the OBOR initiative? To answer these questions, the chapter traces the development of the Port of Shanghai, especially its environmental policies and green development. Also, some recommendations are given to help the Port of Shanghai strengthen its green development and become a 'green' leader on the new Maritime Silk Road.

Keywords: Green port, Maritime Silk Road, Port of Shanghai

5.1 Introduction

Put forward by Chinese President Xi Jinping in 2013, the Silk Road Economic Belt and the 21st Century Maritime Silk Road (B&R) is now a worldwide well-known initiative. Seventy-six major deliverables comprising more than 270 concrete results have been listed in the Belt and Road Forum in May 2017. As President Xi's keynote speech in the forum opening ceremony showed, the fruitful outcomes in the 2013 to 2017 period have deepened policy connectivity, enhanced infrastructure connectivity, increased trade connectivity and strengthened people-to-people connectivity [1]. With the deepening and further development of the B&R initiative, new plans have been gradually published, not only focusing on economic outputs but also on ecological and environmental cooperation [2], energy cooperation [3], green development [4] and so on, which show the emphasis on the B&R environmental impacts.

On 20 June 2017, the *Vision for Maritime Cooperation under the Belt and Road Initiative* was issued by the National Development and Reform Commission and the State Oceanic Administration, demonstrating the effort made to strengthen the Blue Partnership. As is mentioned in the initiative, maritime cooperation and development are experiencing increased focus. Ports, as the connection points on the Maritime Silk Road, are the key to a 'blue engine.' As the worldwide largest container port for the last six years, the Port of Shanghai plays an important role along the Silk Road. Geographical location, economic development, advanced science and technology and inclusive culture all help the Port of Shanghai in its firm determination to become a leader on the Maritime Silk Road.

To set a good example for the ports along the B&R, the Port of Shanghai is actively promoting 'green' port development. In 2015, Shanghai proposed the Three-Year Green Port Action Plan (2015–2017). The Plan is based on energy conservation and emission reduction and emphasizes air pollution control in order to stimulate the construction of a green and clean intensive port and achieve the objectives involved in low carbon and green development. The Port of Shanghai has just completed the Plan. Impressive achievements have resulted from new energy and other technologies, such as, for example, onshore power and LNG technologies, which have reduced polluting emissions sharply.

The Port of Shanghai has been approaching its green goals at high speed, but the concept of a green port is still new in China. Focusing on extension projects and container capacity growth, Chinese ports need to find their own way to a green path. Learning from Western green ports, studying green development measures and innovating new technologies are all necessary to promote green port growth in China. It is also helpful to set a domestic example of a green port. In this respect, the Port of Shanghai would be the best choice. To help the Port of Shanghai achieve the double aim of establishing a green port and leadership on the Maritime Silk Road, this chapter will explore the possibilities and the related policy recommendations.

The main questions of this contribution are: how can the current green development in the Port of Shanghai be evaluated and how can the Port of Shanghai set an example for the B&R project? To answer these questions, in Section 5.2, the paper traces the state of the art of green development in the Port of Shanghai, with the emphasis on policy support for its green port goal. The future measures and steps towards a full green port to help the Port of Shanghai become a green leader on the Maritime Silk Road are given in Section 5.3.

5.2 The status quo of the Port of Shanghai

5.2.1 Development of the port

Located in the Yangtze River Delta, the Port of Shanghai has been the largest container port in the world since 2010. As shown in Figure 5.1, the number of international containers has climbed rapidly from 1995; in 1999, the annual growth rate even reached 37.51%. After a pause in 2009, the growth trend continued, although with the speed declining. With the help of extension and infrastructure projects, the size of the harbour increased in most years, though with fluctuations. So did the number of berths (Figure 5.2). But they both decreased in 2016, which resulted from new regulations for floating cranes upstream of the Huangpu River. At the end of 2016, the size of the harbour was 109.2 km, while the number of berths was 1152.

Since April 2016, the Port of Shanghai has implemented the first stage of a new discharge control area for ships, which has brought and will bring big improvements in air quality, with a 15% reduction of SO_2 density [5]. But with the expected growth of trade and containers in the Port of Shanghai, it is necessary to further control port emissions.

5.2.2 Policy support for green development in the Port of Shanghai

To be a green port, the Port of Shanghai has received policy support from the central government, municipal government, port companies and some collaborating ports. On the national level, some improvements can be seen. Statistics show that in 2016, 5.0 kgce energy has been consumed per 1000 tonnes per nautical mile in ocean and coast freight companies, which is a 4.9%

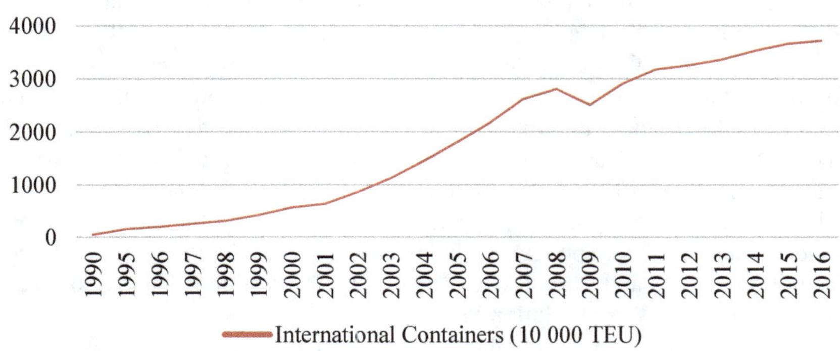

Figure 5.1: International containers throughput in different years. (Source: Shanghai Statistical Yearbook 2017).

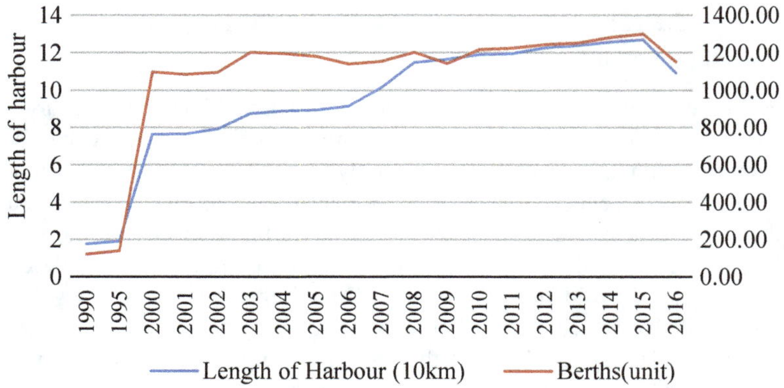

Figure 5.2: Length of harbour and berths in different years. (Source: Shanghai Statistical Yearbook 2017).

decrease compared with the year before. The port companies expended 2.5 tce per 10,000 tonnes, with a reduction rate of 3.0% [6]. In the same year, the investment for waterway transportation environment protection was 5.5 billion RMB, of which 0.72 billion RMB was for ecological protection measures and 3.77 billion RMB was for pollution prevention [6]. Although pollution is declining at a slow rate, there is still a long way to go towards the green port goal. In the Port of Shanghai in 2014, emissions included nitrogen oxide (NO_X) 3.7 Gg, VOC_S 0.5 Gg, CO 1.1 Gg, $PM_{2.5}$ 0.17 Gg and SO_2 0.00728 Gg [7].

To achieve green development, increasingly relevant policies have been issued at the central level. A typical policy is Port Onshore Power Layout Planning [8], in which onshore power is encouraged and regulated. The implementation of the planning achieved emission reductions by 60 Gg SO_2, 110 Gg NO_X and 8Gg $PM_{2.5}$ during peak periods [9]. The Discharge Standard for Water Pollutants from Ships [10] was announced in January 2018 and has been put into effect from July 2018. It was the first time that the regulation involved was amended since 1983. With limitations of different emission sources, the standard provides an important reference for the emissions of ships and makes the regulation and management of water pollutants from ships easier and clearer. More policies include the Special Action Planning for Anti-Pollution of Ships and Ports (2015–2020) [11], Air Pollution Prevention Act Plan [12] and the Water Pollution Prevention Act Plan [13].

At the municipality level, the Port of Shanghai has always taken the lead in regulating port-related environmental impacts. Besides the three-year plan mentioned previously, the Port of Shanghai has introduced policies to achieve green development. Onshore power is a good example. Phase

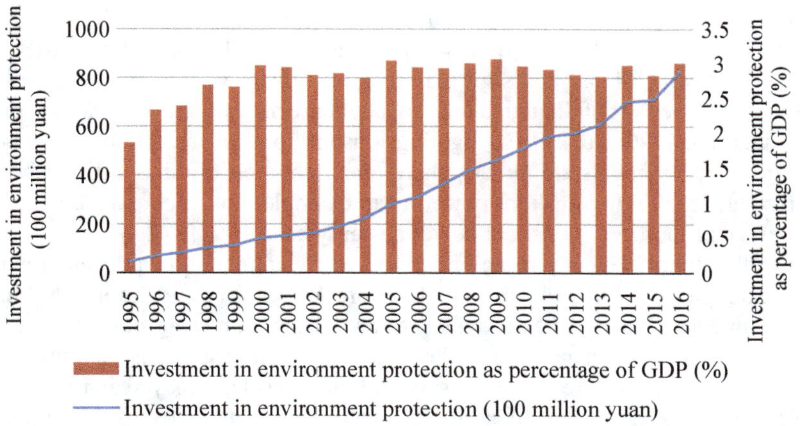

Figure 5.3: Investment in environment protection in different years. (Source: Shanghai Statistical Yearbook 2017).

one of an onshore power project in Shanghai, WuSong International Cruise Terminal, was put into operation in July 2016, which is the largest frequency conversion cruise onshore power system in the world, and also the first one in Asia [14]. On 27 August 2018, the Announcement in Terms of Early Implementation of Emission Control Measurement for Underway Ships in Shanghai Port [15] was issued by the Shanghai Maritime Safety Administration of the People's Republic of China and the Shanghai City administration. Figure 5.3 illustrates the investment in environmental protection of Shanghai city. The investment amount keeps growing throughout the years, while as a percentage of Shanghai's GDP, it first increases and then fluctuates. It shows the attention that Shanghai municipality has paid to environmental protection, and, consequently, a cleaner future can be expected.

5.3 Steps towards green leadership

Although some achievements have been reached as mentioned previously, there is still a lot that can or should be done to become a green leader. Some recommendations are as follows.

5.3.1 Technology and policy innovation

To achieve zero discharge, increase energy efficiency or build a cleaner port, technology innovation is always part of the answer. Although some improvements towards green ports have been achieved, there is still a long way to go. The research and development investments of Shanghai International

Port Co., Ltd. (SIPG) have increased in the past few years, from 24.55 million RMB in 2015, to 37.95 million RMB in 2016 and 45.18 million RMB in 2017 [16], which is a good sign that new and advanced technologies can be expected in the Port of Shanghai to bring about green development. Innovations included the LNG truck, LED lighting, onshore power and so on. Although exciting achievements are already there, the green port aim has still not been fully reached. Taking onshore power as an example, until February 2018, the coverage proportion of onshore power has only been 10% [17].

Policy innovation is also an effective way to simulate green development. For example, the Pilot Planning of Onshore Power for International Sailing Ships in Shanghai Port has provided a reward for promoting onshore power development. But financial subsidies are not enough, and some innovative mandatory measures are also necessary. Policy support has played an important role in helping the Port of Shanghai grow on a green and clean path. And we believe that more policy innovations can come into effect.

5.3.2 Combining green ports with smart ports

Besides a green port, the Port of Shanghai also wishes to be a smart port. The Chairman of SIPG, Chen Wuyuan, said that the main development objectives of SIPG are a smart port, a green port, a technological port and an efficient port [16]. In December 2017, the automated terminal of Yangshan deep-water port began pilot production.

Combining green and smart technologies together would be effective for the Port of Shanghai. The smart port goal can provide advanced technologies to increase production and manufacturing efficiency. Higher energy efficiency and lower energy consumption can be significant aims of a smart port, which is closely related to a green port. On the other hand, a green port cannot be realized without energy consumption reduction. For instance, artificial intelligence can change the construction of an onshore power facility in a positive way [18]. In Xiamen port, the smart logistics system has improved the current transport system, reduced the waiting time for order receipt and decreased emissions from transportation automobiles [19]. Therefore, with the same ambitions, green and smart should have mutual promotion.

5.3.3 Strengthening cooperation with other green ports

Cooperation is a good way to learn from and move forward with other ports. We already argued that strengthening cooperation with other cities can help Shanghai city [20]. This also applies to the port. Friendly cooperation between Gothenburg port and the Port of Shanghai has been ongoing since 1994. The relationship promotes mutual trade and technology and staff exchange, which speeds up the development of both ports.

Under the Belt and Road Initiative, more cooperation opportunities are being provided. It is a valuable platform for the ports along the Silk Road. Cooperation boosts the development of participating ports and their countries, which is a win–win situation. Not only economic convenience can be achieved but Sustainable Development experiences can also be exchanged. For example, Dong Ma, the project manager of the Vehicle Emission Control Centre, claimed that though some achievements have been reached, such as setting emission standards and building emission control areas, there is still no ship and port emission inventory being done [21]. So, it would be extremely helpful to learn from databases built in other ports.

5.4 Conclusions

The Port of Shanghai is a leader in terms of containers being handled and is on its way towards being a green leader [22]. We analysed the development status of the Port of Shanghai and evaluated the policy environment. The results show that the port has made a lot of progress in the past few years towards green development, but there is still a lot to be done for green leadership. To help Shanghai achieve its green port aim, we recommend promoting innovation in technology and policy, combining green port and smart port objectives and strengthening cooperation with other ports along the Maritime Silk Road.

References

[1] Full text of President Xi's speech at opening of Belt and Road forum, Xinhua Net. http://news.xinhuanet.com/english/2017-05/14/c_136282982.htm. Accessed on: 1 December 2017.

[2] MEP, The Belt and Road Ecological and Environmental Cooperation Plan. Beijing: Protection MoE (ed.), 2017.

[3] NEA, NDRC, Energy Vision and Actions on Jointly Building Silk Road Economic Belt and 21st Century Maritime Silk Road (in Chinese). Beijing: NEA, NDRC, 2017.

[4] Protection MoE, Affairs MoF, Commission NDaR, Commerce Mo (eds.), MEP, MFA, NDRC, MFC, Guidance on the Formation of Green 'Belt and Road', 2017.

[5] Wang, J., The three-year act plan goals of green port for Shanghai have been realized basically. http://sh.eastday.com/m/20171031/u1ai10962949.html. Accessed on: 5 May 2018.

[6] China Ports and Harbours Association. China Ports Year Book 2017 (in Chinese). China Ports and Harbours Association, 2017.

[7] Wen, P., Chinese ports anti-pollution should speed up towards the right lines (in Chinese). *China Ports*, pp. 19–21, 2015.

[8] Ministry of Transport (ed.), Water Transportation, Port Onshore Power Layout Planning. Beijing: Ministry of Transport, 2017.

[9] Interpretation of 'Port Shore Power Layout Plan'. http://web.archive.org/web/20180226080335/http://zizhan.mot.gov.cn/zfxxgk/bnssj/syj/201708/t20170801_2799499.html. Accessed on: 2 July 2021.

[10] Ministry of Environmental Protection, AQSIQ (eds.), Discharge Standard for Water Pollutants from Ships. Beijing: Ministry of Environmental Protection, 2018.

[11] Water Transportation Bureau, Special Action Planning for Anti-Pollution of Ships and Ports, Transport Mo. Beijing: Water Transportation Bureau, 2015.

[12] The State Council, Air Pollution Prevention Act Plan, 2013.

[13] The State Council, Water Pollution Prevention Act Plan, 2015.

[14] Li, Y., The world's biggest cruise onshore electricity system put into operation at Wusong Port to achieve zero discharge at the port. https://finance.ifeng.com/a/20160717/14606332_0.shtml. Accessed on: 2 July 2018.

[15] Xinde Marine News, Shanghai Port implements stricter emission control on Oct. 1. https://www.xindemarinenews.com/m/view.php?aid=6856. Accessed on: 2 July 2021.

[16] Technology innovation in Shanghai port, *China Ports*. http://www.chinaports.com/portlspnews/%205875b708-d369-4719-9b5b-619eb8fbcad0. Accessed on: 20 September 2018.

[17] The approach of forging onshore electricity for 'green shipping' is worth popularizing, *Jiefang Daily*. http://sh.eastday.com/m/20180606/u1ai11498209.html. Accessed on: 1 October 2018.

[18] Han, X. & Shou, J., The development and prospect of onshore power construction in China's green ports (in Chinese). *Modern Enterprise*, pp. 49–50, 2018.

[19] Huang, Z., The construction and reflection for green port system in Xiamen port. *China Ports*, 2018.

[20] Li, Z., Galván, M.J.G., Ravesteijn, W. & Qi, Z., Towards low carbon based economic development: Shanghai as a C40 city. *Science of the Total Environment*, 576, pp. 538–548, 2017.

[21] Zhang, K., Ship and port emission has been one of the important air pollution sources. https://www.yicai.com/news/5420991.html. Accessed on: 15 September 2018.

[22] Li, H. & Lv, S., Estimation of cruise environmental pollution and analysis on control scenario: A case study in Shanghai (in Chinese). *Ocean Development and Management*, 33, pp. 32–38, 2016.

CHAPTER 6

Management of large technological projects: The case of Gwadar Port in Pakistan

Dimosthenis Michalakeas[1] & Oscar Garay Garcia[2]

[1]GAIAOSE SA, Greece

[2]South America Energy Business Development,
Schneider Electric Energy and Sustainability Services, Chile

Abstract

Large infrastructure projects such as the Gwadar Port project in Pakistan are frequently deprived of design requirements with respect to public engagement and long-term orientation. Plans are mainly focused on the economic benefits of the forthcoming infrastructure construction and neglect other aspects such as environmental and social impacts. As a result, these projects can face a lot of opposition that may threaten their realization. In order to avoid these negative consequences, a complete plan should be designed in advance, taking into consideration the notion of Responsible Innovation. This idea implies the acquisition of knowledge of the consequences of possible actions, their evaluation and, finally, the utilization of this information as a functional input to the project. What is more, such plans will enable all people who are influenced by the decision-making processes to be part of the design and development. Consequently, these people have the opportunity to state their objections and concerns that ultimately could be included in the final decision and implementation. Additionally, it will allow involved members to alter or adjust plans and respond to emerging issues without 'excess' costs. The present chapter extensively analyses the case of the construction of Gwadar Port. It identifies project complexities, value conflicts among stakeholders and a variety of possible impacts that pertain to society, the environment and the economy. Following this analysis, possible project improvements that will enhance its overall performance, public approval and sustainability are suggested. Based on these improvements, a more efficient and effective implementation plan is proposed. The main conclusion this chapter gives is that not only the Responsible Innovation principles were not followed in the Gwadar Port project, but also some ethical principles. To fix those shortcomings, a seven-step implementation plan is designed in order to include Responsible Innovation and ethical principles into the Gwadar Port project.

Keywords: Gwadar Port, Responsible Innovation, value conflicts, implementation plan, impact assessment

6.1 Introduction

The globalizing world, nowadays, dictates governments and international firms to minimize operational costs and time-consuming functional procedures in order to make more profit, tighten relationships among neighbouring countries or both. To achieve these targets, cooperation among countries is essential. This usually takes the form of a mega project for which financial support and mutual engagement are needed. The construction of the Gwadar Port constitutes an obvious example of such a relation.

Since mega projects are expected to last for many decades, a good design is considered to be critically important. Thus, the Sustainable Development of a project should not only consider economic factors but also meet long-term targets. This long-term future vision implies (according to the definition of sustainability) that the development of an infrastructure project should be designed in a way that will not compromise the needs of future generations, enabling them to exploit the infrastructure according to their own needs.

As a result, Responsible Innovation plays (or should play) an important role in designing infrastructure. According to Van den Hoven et al. [1], Responsible Innovation (RI) refers to ways of proceeding in research and innovation that allow those who initiate and are involved in these processes at an early stage to:

A. Obtain relevant knowledge on the consequences of the outcomes of their actions and the range of options open to them
B. Effectively evaluate both outcomes and options in terms of moral values (including, but not limited to well-being, justice, equality, privacy, autonomy, safety, security, sustainability, accountability, democracy and efficiency) and perspicuously represent value commitments and discuss them with relevant stakeholders
C. Use these considerations (under A and B) as functional requirements for the design and development of new research, products, infrastructure, systems and services

In addition to the definition of Van den Hoven et al. [1], Ravesteijn et al. [2] propose a five-step procedure to operationalize the Responsible Innovation concept:

1. Stakeholder analysis
2. Impact assessment
3. Identify diverse values of stakeholders
4. Analyse value conflicts among stakeholders
5. Search for value conflicts solutions

The objective of the present chapter is to make a proposal to improve the collaborative (China and Pakistan) infrastructure construction of the Gwadar Port and subsequently to design a more efficient and effective implementation plan in which the fundamental aspects of Responsible Innovation have been considered. In order to do so, the five steps that Ravesteijn et al. [2] suggested will be followed. This proposal bears the character of an advisory report for the China National Development and Reform Commission (CNDRC), the main stakeholder from the Chinese government side that aims to develop Gwadar Port. It aims to help the CNDRC incorporate some Responsible Innovation principles that may be missing.

Section 6.2 starts with an analytic description of the Gwadar Port and, subsequently, project complexities are analysed in Section 6.3, especially technical, organizational and social complexities. For the latter type, a more comprehensive examination is conducted, including a stakeholder analysis and a value analysis. The potential impacts from Gwadar Port are scrutinized in Section 6.4. These impacts are appraised with respect to social, environmental, economic and safety aspects. Consequently, in Section 6.5, value conflicts between stakeholders are analysed and some solutions are proposed. Section 6.6 proposes potential and possible improvements for the whole project, while section 6.7 suggests an implementation plan, in which all these improvements have been incorporated. Finally, the chapter concludes with a summary of the whole study.

6.2 Gwadar Port description

Gwadar Port is a warm-water, deep-seaport on the shores of the Arabian Sea that is located in the southwest part of Pakistan in Balochistan Province (Gwadar City) about 120 km from the Iranian border (see Figure 6.1). It is the third-largest deep-seaport of Pakistan after Karachi and Port Qasim.

6.2.1 Port construction

The construction of the present port consisted of two phases. Phase I started in 2002 and lasted until December 2006, costing around US$248 million. Phase II, still underway, started in 2007; it is estimated to cost about US$932 million. These two phases include the following [3]:

Phase I

- Berths: three multipurpose berths
- Capacity: bulk carriers of 30,000 deadweight tonnage (DWT) and container vessels of 25,000 DWT
- Length of berths: 602 metres
- Approach channel: 4.5 km long dredged to 12.5 m depth

Figure 6.1: The location of Gwadar.

- Turning basin: 450 m in diameter
- Service berth: one 100 m berth
- Related port infrastructure and handling equipment, pilot boats, tugs, survey vessels, etc.
- Type of ships: bulk carriers of 30,000 DWT and container vessels of 25,000 DWT

Phase II

- Four container berths
- One bulk cargo terminal (capacity: 100,000 DWT ships)
- One grain terminal
- One ro-ro terminal
- Two oil terminals (capacity: 200,000 DWT ships each)
- An approach channel to be dredged to 14.5 m depth
- Type of ships: 200,000 DWT vessels

After the completion of construction phase I, the Pakistani government and the Chinese company known as China Overseas Port Holding Company (COPHC), on 1 February 1 2007, agreed to a 40-year lease contract to PSA (Port of Singapore Authority) International Pte Ltd for the operation and development of the Gwadar Port. However, in January 2015, the Chinese Overseas Holdings Limited took over control with the consent of Pakistan [4]. Thus, although the port belongs to the government-owned Gwadar Port Authority, it will be operated by the state-run Chinese company. The area assigned to the Chinese company covers more than 2000 acres and, according to the agreement, tax-free port and duty-free trades zones are included. Since this transfer, a series

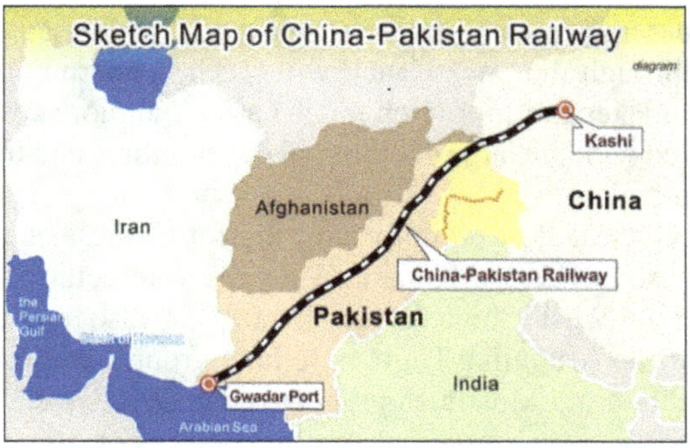

Figure 6.2: China–Pakistan Economic Corridor (CPEC).

of major expansions including airports, hospitals and buildings have also been undertaken to turn Gwadar into a full-fledged, deep-water commercial port.

What is more, in order to strengthen and tighten their relationship, in 2013, the Chinese and Pakistani government agreed to build the China–Pakistan Economic Corridor (CPEC). The CPEC is a vital link for China because it will connect Gwadar Port and the heartland of China, and as a result, it will be used to supply merchandise to China's mainland (see Figure 6.2). Although this economic corridor is an independent project, it is considered to be under the umbrella of the whole cooperation between the two countries. Moreover, the CPEC project is not only limited to Gwadar Port infrastructure construction since it will also boost energy cooperation and involve other potential investment programmes. In the meantime, the China–Pakistan railway is under construction.

6.2.2 Strategic importance

Gwadar Port has significant economic and strategic importance. The port is adjacent to the entrance of the Persian Gulf, which is a gateway for about 20% of the world's oil. Consequently, the port lies on a major route for tankers carrying oil around the world.

It is obvious that the port, together with the construction of the new corridor, will provide a crucial connection between China and the Persian Gulf as well as Africa, Europe and other parts of the world. Cargo and oil from (and to) Chinese cities will no longer use the existing maritime routes through the Strait of Malacca and around India (Figure 6.3). They will rather follow a detour route taking advantage of the mainland of both China and Pakistan by using railways and highways, such as the Karakoram Highway, connecting China with Pakistan, which the Chinese government has planned to upgrade in order to pursue long-term profits [5]. Consequently, Chinese cargo, passing through the Suez Canal, will reach Western countries quicker. This connection makes transportation much easier and more cost-effective and efficient compared with the large operational expenditures and time-consuming present routes of tankers.

In addition, Gwadar Port is very promising for Chinese economic growth, especially in the southwestern Xinjiang Province (prefecture of Kashgar, see Figure 6.3) and in West China as a whole. This also provides economic opportunities for the struggling Uighurs (ethnic group), while it can improve relations of neighbouring Muslims in the two countries as well.

As far as Pakistan is concerned, the infrastructure construction will largely benefit the strategic importance of the country. Currently, Pakistan does not have the power to influence strategic decisions in its vicinity since it is considered to be one of the poorest countries in the world [6]. However, with Gwadar Port, Pakistan will be a key player in the region. The port will

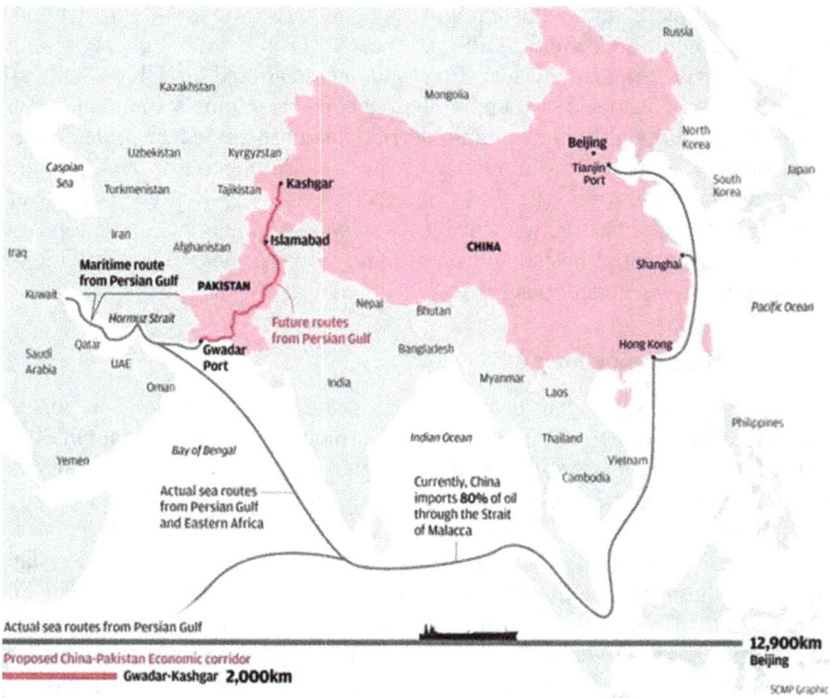

Figure 6.3: China-Pakistan Economic Corridor (CPEC).

also potentially benefit the economy of Pakistan, in particular, the city of Gwadar and the province of Balochistan. It is estimated that, from construction to operation, the port will create approximately two million jobs. Moreover, urbanization, increased trade and commerce as well as industrialization are expected to provide additional direct benefits for Gwadar Port in the short and long term, resulting from and giving a boost to economic growth and development.

6.2.3 Choice of exact location

As can be seen in Figure 6.3, the route through Pakistan constitutes the shortest way to the Persian Gulf and thus the location of Gwadar Port is significant [7]. The specific place of Gwadar Port was selected for the following reasons:

- Gwadar Port is considered to be a gateway to the Persian Gulf at the Strait of Hormuz (Pakistan already has two large operational ports, but not so far west)
- Its location (at the mouth of the Persian Gulf) potentially enables more harbour services, such as storage of sea resources as well as shipment, trans-shipment and manufacturing conveniences for regional and non-regional key players (like the United Arab Emirates and the European Union)
- Its location is close to the international sea lines of communication (SLOC) and it is suitable for trans-shipment facilities
- It offers protection against natural disasters
- It will potentially contribute to the development of the heartland

6.3 Project complexities

Large infrastructure projects, such as Gwadar Port, face complex and complicated decision-making processes not only because of inherent complexities but also because of stakeholder disagreements. Complexities can make projects fail, and hence a thorough analysis of possible complexities must be carried out for every large infrastructure project. For such an analysis of Gwadar Port, the Technology–Organization–Environment (TOE) framework for project complexities is applied [8]. The TOE framework, distinguishing internal and external complexities, provides a method for assessing and managing project complexities in order to improve the performance of large engineering projects.

6.3.1 Internal complexities

According to Bosch's TOE framework [7], internal complexities consist of technical (T) and organizational (O) complexities. Technical complexities relate to the goals, tasks and scope of the project. Organizational complexities have to do with the necessary resources for the project, the working teams and the trust between them.

6.3.1.1 Technical complexities

The TOE framework uses 17 criteria for technical project complexity. Each of these was given an ordinal scale assessment (low, medium, high), which, according to rough estimations, resulted in Table 6.1.

Although one might think that in an ambitious project like Gwadar Port the technical complexities would be considerable, however, they are relatively limited. During the past 50 years, the Chinese government and companies gained a lot of experience in port construction, and Chinese engineers involved in port construction are highly skilled and capable to determine clear and feasible project objectives. Nevertheless, some criteria score high, such as the variety and number of tasks, which calls for special attention. Generally, the technical complexities can be effectively addressed through continuous efforts and innovations.

Table 6.1: Technical complexities and complexity level.

Technical complexity	Complexity level
High number of project goals	High
Non-alignment of project goals	Low
Unclarity of project goals	Low
Uncertainties in scope	Medium
Strict quality requirements	Medium
Project duration	High
Size in CAPEX	High
Number of locations	Low
Newness of technology	Medium
Lack of experience with technology	Low
High number of tasks	High
High variety of tasks	High
Dependencies between tasks	High
Uncertainty in methods	Low
Involvement of different disciplines	High
Conflicting norms and standards	Low
Technical risks	Low

6.3.1.2 Organizational complexities
During the construction of Gwadar Port, interaction and cooperation between individuals and organizations from the two countries are inevitable and crucial. So far, the progress of the Gwadar Port project has been relatively smooth and successful. However, even though the friendship between China and Pakistan is quite strong, differences in languages, culture, legislation and education result in an extremely complex situation for project management, as shown in Table 6.2 (made in the same way as Table 6.1).

Organizational complexities can be overcome through better communication. For example, in order to improve Chinese workers' affiliation with local citizens and guarantee their safety, continuous communication is necessary not only at the national level but also at the regional and organizational levels. Certainly, additional actions and policies will be required for the successful execution of the project.

6.3.2 External complexities

The last element of the TOE framework, environmental (E), refers to the external or social-environmental complexity and is analysed in this subsection.

Table 6.2: Organizational complexities and complexity level.

Organizational complexity	Complexity level
High project schedule drive	High
Lack of resource and skill availability	High
Lack of experience with parties involved	High
Lack of HSSE awareness	High
Interfaces between different disciplines	High
Number of financial sources	High
Number of contracts	High
Variety of contracts	High
Number of different nationalities	High
Number of different languages	High
Presence of joint venture partners	Medium
Involvement of different time zones	High
Size of project team	High
Incompatibility between different PM methods/tools	High
Lack of trust in project team	Medium
Lack of trust in contractor	Medium
Organizational risks	High

This complexity includes stakeholders, location, market conditions and risks [7]. Identifying the stakeholders and analysing their positions plays an important role since the parties involved can influence the decision-making processes in various ways [9]. In this section, first, a stakeholder analysis is conducted and, second, fundamental values of these actors are explored.

6.3.2.1 Stakeholder analysis

Given that a large number of actors are involved in this project, it is not possible to include all parties in the analysis and thus we focus on critical actors. A stakeholder is considered to be any group or individual who has an interest or is affected by the consequences of the project or can influence the solutions that are considered [8]. Moreover, to clarify the interdependencies of the actors, three power positions are distinguished: production power, blocking power and diffuse power [10]. The parties that positively influence Gwadar Port construction are considered to have production power. Actors with an interest in preventing construction have blocking power. Additionally, some stakeholders have diffuse power, which is the combination of production and blocking power. The result of the stakeholder analysis is shown in Table 6.3.

Gwadar Port is a critical port for the Chinese One Belt, One Road strategy, which aims to develop the Chinese economy through the Silk Road Economic Belt and the 21st Century Maritime Silk Road [11]. Additionally, the Chinese

Table 6.3: Stakeholder analysis: interests, power, goals, existing and desired situations.

Stakeholder	Interests	Goals	Existing situation	Desired situation
China National Development and Reform Commission (CNDRC)	Economic development, political influence	Build economic corridor between China and Western countries; higher GDP; better transportation routes; more port users; better infrastructure support; safety	Instability in Pakistan; high cost in building port; out-of-date infrastructure in Pakistan	Safety of employees; new technology to reduce construction cost

(Continued)

Table 6.3: Continued.

Stakeholder	Interests	Goals	Existing situation	Desired situation
Chinese Petroleum Importers (CPI)	Profit in petroleum imports	More profits; easier access to petrol source; safety	High cost of petroleum import transportation; poor geographical conditions	Lower cost of petroleum import transportation; regional stability
Petroleum Exporters in Persian Gulf (PEPG)	Profit in petroleum exports	To attract petrol buyers; more profits; easier access to petrol source; safety	Long-distance of petroleum transportation	Direct petrol delivering pipelines
China Overseas Holdings Limited (COHL)	Overseas ports	To have more overseas ports; more profits; more port users; safety	Poor regional infrastructure support; instability in Pakistan; lack of engineers in Pakistan	More investments in local infrastructures; social stability
Pakistan Central Government (PCG)	Economic development	To attract foreign investments in Pakistan; higher GDP; more taxes; more job opportunities; safety	Social instability; high unemployed rate; slow economic growth	More job opportunities; social stability; more foreign investments; rapid economic growth
Committee on Economic Corridor between China and Pakistan (CECCP)	Trade between China and Pakistan	To promote trade between China and Pakistan; higher GDP; better transportation routes; safety	Poor infrastructure support in Pakistan; safety	More investments in infrastructures; social stability
China Railway Construction Corporation (CRCC)	Profit in railway construction	To build a railway between China and Pakistan; more profit; deliver on time; safety	Poor geographical conditions; social instability	New technology to support construction; safety

(Continued)

Table 6.3: Continued.

Stakeholder	Interests	Goals	Existing situation	Desired situation
Other Import and Export Corporations (OIEC)	Profit in import and export trades	To reduce transportation costs by using Gwadar Port; more profit; safety	Social instability; poor infrastructure support	Low cost in transportation; safety
European Countries (EC)	Profit in import and export trades	To reduce transportation cost by using Gwadar Port; higher GDP; easier access for imports and exports	Social instability; poor infrastructure support	Low cost in transportation; safety
PSA International Pte Ltd (PSAIPL)	To maintain the high position of Singapore Port	To control Gwadar Port; more profit; dominating position in ports	Low profit in Gwadar Port; social instability	Increased profit in Singapore Port; Higher profit in Gwadar Port
United States (USA)	Political rights in South Asia	To prevent China from having the controlling power of Gwadar Port; military control; higher GDP	Weak influence in South Asia	Powerful rights in the construction, transportation of South Asia
India (I)	Economic development	To build economic corridor between India and China; higher GDP; better imports and exports	Conflict between China and India	Increased trade between India and China
Citizens of Gwadar (CG)	Living conditions	More job opportunities; safety; better water quality; less noise; better air quality	Social instability; few job opportunities; poor infrastructure; poor education	Social stability; more job opportunities; more investments in infrastructure

Note: Red means high power or strong interests, green low power or weak interests.

government also tries to restrain US political influence in South Asia. The CNDRC aims to build an economic corridor between China and European countries through Gwadar. The CNDRC represents the central Chinese government and has both high power and strong interest in Gwadar Port. However, Gwadar cannot provide good labour or a safe working environment for the operation [12]. Also, a railway is still needed to transport goods from Gwadar Port to China, otherwise, the port cannot even support its own economic development; the railway, however, is under construction.

Import and export enterprises are also interested in the port to minimize transportation costs, especially petroleum companies [13]. PSA, the previous manager of Gwadar Port opposes the operation of the port to protect its interests in Singapore, another alternative for import and export enterprises. For the United States, Gwadar Port has both economic and political importance due to its location and, consequently, the United States also tried to obtain control. The Indian government has argued over border issues with China, but at the same time, they would like to build an economic corridor between China and India (instead of Pakistan). The citizens of Gwadar City are another important actor involved in the port. On the one hand, they will welcome more job opportunities and a boost to the local economy, but, on the other hand, and because of the lack of education in the area, locals still look at this mega project as an enemy to their traditional way of living.

In order to visualize the interdependencies in Table 6.3, a power–interest matrix has been made based on interests and power positions (see Figure 6.4). China Overseas Holdings Limited, the Pakistan central government, the Committee for the economic corridor between China and Pakistan and the citizens of Gwadar are the key actors that the CNDRC should consider when making decisions or operating the port. China Railway Construction Corporation, the United States and India should be kept informed about port developments as they have high interest but low power.

6.3.2.2 Value analysis

The identification of stakeholder values is an important next step in a Responsible Innovation approach. Following Ravesteijn et al. [2], for this purpose, first, the stakeholder goals (in Table 6.3) can be divided into one of the three P (People–Profit–Planet) categories and, second, the main underlying values can be determined (see Tables 6.4 and 6.5).

6.3.2.3 Two value trees

These main values can be further specified. If all these specified values are grouped together, a value tree can be constructed, representing a hierarchy of values [14]. We will do so for two important stakeholders (see Figures 6.5 and 6.6).

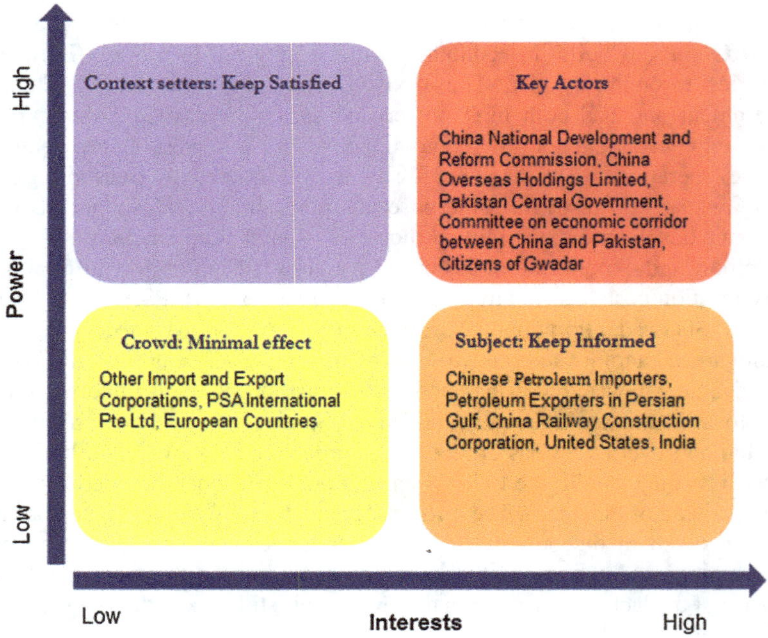

Figure 6.4: Stakeholder analysis power–interest matrix.

Important values for the CNDRC are:

- Safety and security. Pakistan is not a stable country and severe conflicts frequently occur. Thus, safety first, including the security of employees.
- Sustainability. Sustainable goals and objectives should be put high in the value tree, not only to secure long-term port operation but also to ensure that future generations in Gwadar will have the same or better living conditions than today. The port's clients and transportation companies are the major actors here.
- Economic. Economic benefits are crucial to the China National Development and Reform Committee. Among various industries, petroleum is the most important one as the port is quite close to oil resources in the area. Capacity and maintenance are critical factors here.
- Political. Besides China and Pakistan, other countries are also involved, such as the United States and India. By controlling the port, China strengthens its (military) influence in South Asia.

- Reputation. In the long run, China could improve its reputation by supporting Pakistan and other surrounding poor countries in their economic growth and development.

Important values for the citizens of Gwadar are:

- Safety and security. The inhabitants of Gwadar are very concerned about their safety and security because of the political and religious conflicts in Pakistan.

Table 6.4: Classification of stakeholder goals.

Stakeholder	Economic	Environmental	Social
CNDRC	—Higher GDP		—Better infrastructure support
	—Better transportation routes		—Safety
	—More port users		
CPI	—More profits		—Safety
	—Easier access to petrol source		
PEPG	—More profits		—Safety
	—Easier access to petrol source		
POHL	—More profits		—Safety
	—More port users		
PCG	—Higher GDP		—More job opportunities
	—More tax		—Safety
CECCP	—Higher GDP		—Safety
	—Better transportation routes		
CRCC	—More profit		—Safety
	—Deliver on time		
OIEC	—More profit		—Safety
PSAIPL	—More profit		
	—Higher dominating position in port		

(Continued)

Table 6.4: Continued.

Stakeholder	Economic	Environmental	Social
US	—Military control —Higher GDP		
EC	—Higher GDP —Easier access for import and export		
I	—Higher GDP —Better import and export		
CG		—Better water quality —Better air quality —Less noise	—More job opportunities —Safety

Table 6.5: Value analysis of stakeholder goals.

Stakeholder	Economic	Environmental	Social
CNDRC	Economic growth		Social stability
CPI	Profitability		Social stability
PEPG	Profitability Accessibility		Social stability
POHL	Profitability Productivity		Social stability
PCG	Economic growth		Social stability Liveability
CECCP	Economic growth Accessibility		Social stability
CRCC	Profitability Productivity		Social stability
OIEC	Profitability		Social stability
PSAIPL	Profitability		
US	Economic growth Strategic and political		
EC	Economic growth Accessibility		

(Continued)

Table 6.5: Continued.

Stakeholder	Economic	Environmental	Social
I	Economic growth		
CG		Environmental friendliness	Liveability Social developments

- Environmental friendliness. The scarcity of drinking water constitutes one of the major challenges. Without a constant flow of pure water, the social development of the area is very unlikely. In addition, due to port construction and operation, the air quality might be reduced, not only by the port itself, but also by the inflow of increasing numbers of people.
- Liveability. If new and more jobs are available, people's living conditions will improve due to higher household incomes. However, there are also jobs lost with land loss and reduced fish stocks. Also, respect for their traditions is crucial to the citizens.
- Social development. Equal and good education and healthcare to meet social basic needs, and strengthen the economy as well.

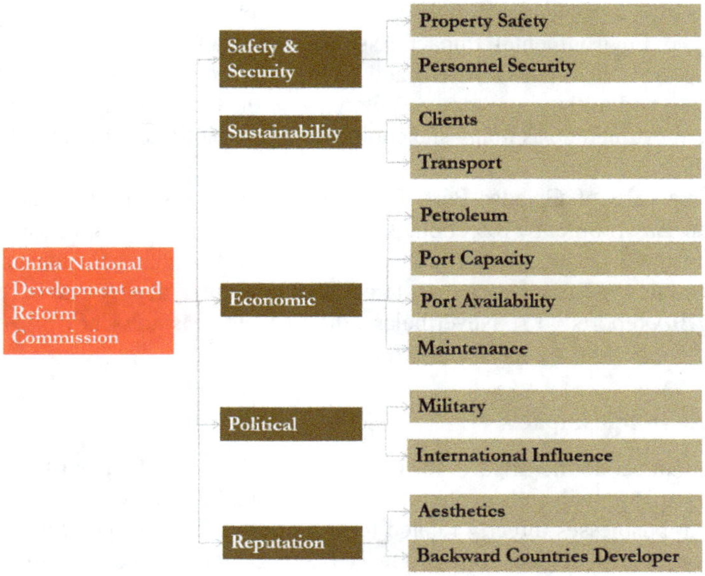

Figure 6.5: Value tree for CNDRC.

Figure 6.6: Value tree for CG.

6.4 Impact assessment

One of the most important parts of an infrastructure or technological project is to perform an extensive impact assessment research. Social Impact Assessment (SIA) and Environmental Impact Assessment (EIA) are the most common and important methods. Positive and negative effects have to be identified, enhancing and mitigating them, respectively, in the course of the project. This process must include as many stakeholders as possible in order to avoid major conflicts.

In the case of Gwadar Port, Pakistani law requires an EIA study. The Pakistan Environment Protection Agency clearly states on its website: 'All developmental projects from public and private sector that fall under any of the Schedules of Regulations have to obtain environmental approval in respect of IEE/EIA reports' [15]. Nevertheless, the EIA report is not accessible online (for the public). So we have to use other sources!

6.4.1 Economic impacts

Three main economic impacts for China and Pakistan can be identified. First, the economic growth and development of China and Pakistan will be stimulated thanks to businesses directly related to port construction or operation as well as all other sub-services and products that indirectly serve port functions. Ultimately, this might also give citizens better opportunities to increase their standard of living.

Second, Gwadar Port will increase employment not only in Pakistan but also in China. Large infrastructure projects such as highways, power plants, airports or ports require a lot of labour, but also promote employment in the long term when the project result is fully operational and serves as an economic driver for the country. This increase in employment will also have second-order effects, such as higher wages and even lower interest rates for loans, according to macroeconomic theory. Lower interest rates allow more investment in infrastructure projects, increasing employment again.

Third, a positive effect can also be identified for other countries that export/import a great quantity of goods to and from China, such as raw materials and other products. The port will enable products from the EU to reach China (and vice versa) in less time, decreasing total shipping and operational costs. In this sense, the EU will also get a positive second-order effect from Gwadar Port.

Fourth, some countries could be negatively affected by Gwadar Port, especially neighbouring countries that also have important ports. For instance, in India, there are 76 ports similar to Gwadar Port and these will suffer from the fact that almost all Chinese cargo ships will use Gwadar Port instead of other ports in the region. Countries facing such effects are India, Kuwait, Bahrain, Qatar, United Arab Emirates and Oman. Figure 6.7 shows the most important ports in the area [16].

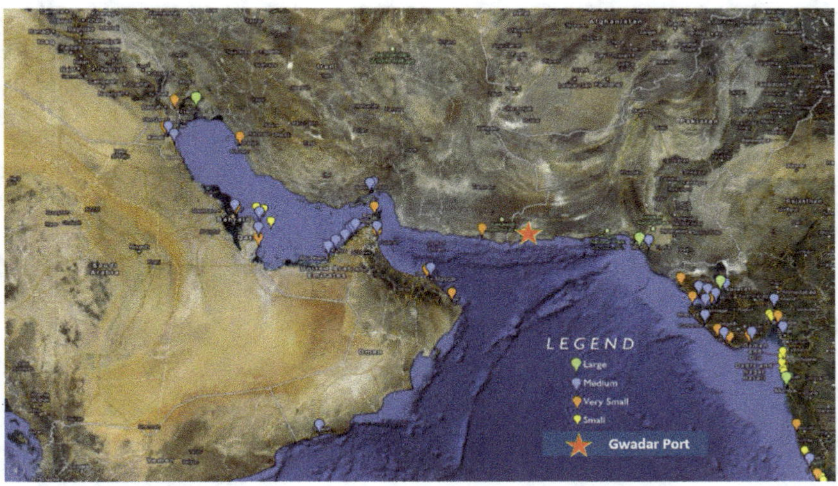

Figure 6.7: Ports of the region.

6.4.2 Social impacts

According to one of the Pakistani governmental documents about Gwadar Port, the main objective of this project is the 'transformation of the district from a marginalized rural society to an economically vibrant industrial and commercial community' [17]. Consequently, major social changes can be expected in the Gwadar area, including the four following changes.

First, due to the fact that the main economic activity in the area is fishing, craft fishermen will suffer the consequences from both construction and operation of the port. Heavy cargo ships going in and out will highly affect the fish stock and the households that rely on it. Consequently, compensation measures would have to be taken in favour of those fishermen who will lose their business.

Second, major demographic changes will occur in the area. The Gwadar area has an extremely low literacy rate (26%) [17]. Because there are no local people with enough knowledge and skills to perform jobs related to port construction, workers will come from elsewhere during this phase. This might have second-order impacts, such as food and water scarcity and shortages in the domains of housing, hospital care and schools. Third-order impacts could include the proliferation of diseases, crime and traffic jams.

Third, in a large project like this, in which the objective is to completely transform a poor area into an international trade and business centre, resettlement of people is always an issue. Especially during construction, people will have to be reallocated to different places in the country. This is a major challenge for Pakistan, in which poverty and terrorism are spread all over the country. In fact, estimates are that 35,000 people have died from terrorism between 2001 and 2011 [18]. For that reason, it is not an easy task just to resettle people. Authorities and project leaders should thoroughly consider the security and living conditions of potential new areas.

Fourth, changes as to gender can be expected. Such a large infrastructure and technological project with ambitious objectives will change the way people look at women when it comes to their responsibilities and capabilities. The whole development would result in higher gender inclusiveness in the local community due to improvements in people's literacy and education. This is a very important impact to consider in a country like Pakistan, in which women have fewer rights and opportunities than men.

6.4.3 Environmental impacts

In the case of such a large infrastructure and technological project, major environmental impacts can be expected. First, mangrove forests can be found in the Gwadar coastal area, specifically in Kalmat Hor and Gwadar Bay. Here, mangrove forests cover more than 4000 hectares [17]. These forests serve as nurseries and breeding grounds for shrimp, fish, crabs and other marine species. As such, they offer a large amount of fauna to the sea, which constitutes the main economic basis of the area in the form of fisheries. Locals strongly protect these forests because they keep the whole sea ecosystem perfectly balanced. Gwadar Port would greatly affect the forests, especially with the construction of the Mirani Dam near the mangrove area, which will serve as a freshwater source.

Second, it is likely that some of the livestock in the region must be resettled to other areas. The construction of a port in a poor and undeveloped area like Gwadar comes with other major infrastructure projects, such as highways, housing and urbanization in general. Due to this, not only coastal zones would be affected, but also inland areas. Inland areas in Gwadar hold around 72,000 livestock animals, like cattle, small ruminants and pack animals [17]. If some of that livestock population is resettled, the environmental balance is likely to change.

Third, with the construction of the port, new roads and highways, noise and air pollution will inevitably rise. Consequently, measures should be taken to keep pollution away from residential areas, such as building soundproof walls close to highways, installing pollution filters to the working machinery in the port and policy changes to keep pollution at minimum levels.

Fourth, the local seabed ecosystem will be affected. Several organisms rely and live on the seabed, such as different seaweed species, shellfish and all other species that feed themselves with the resources there. Sand extraction is the most damaging and, consequently, measures should be taken to reduce this impact as much as possible.

Fifth, agriculture will drastically change. To support a huge inflow of migrants, local agriculture has to be developed to provide enough food for the people. The expected agricultural development could benefit from unused lands. As a result, the whole inland natural environment will change. However, expansion of farming would require a continuous flow of water, which is an extremely scarce resource in this harsh and dry environment.

Sixth, oil spills might jeopardize the whole natural ecosystem. In view of the rich oil resources available in the region, Gwadar Port will likely be used by tankers. A single accident in which tonnes of oil are spilt into the ocean would kill thousands of animals in the area, affecting fisheries as well as the port operations.

6.4.4 Safety impacts

According to the characteristics of Gwadar Port, its geographic location and the political situation in Pakistan, three main impacts can be identified related to safety issues. Given its economic importance, the first and probably the most important issue pertains to a possible terrorist attack aimed at hitting powerful stakeholders (e.g. governments involved) by causing severe damage to the port. If a terrorist group is willing to attack China or Pakistan, Gwadar Port would be the perfect target to realize its mission [17, 19]. Thus, not only the project leaders but also both governments should take this aspect into account when designing the safety measures of the port.

Second, the Mekran coastal zone, in which Gwadar is located, is one of the most seismically active regions in Pakistan, susceptible to earthquakes and tsunamis. Statistics indicate that between 1851 and 1990, a total of 193 earthquakes hit the area with a magnitude of 4.0 or higher on the Richter scale [17]. This threat must be considered carefully when designing emergency protocols and construction criteria for the port in order to reduce possible devastating consequences.

Third, even though the region is very dry with hot summers, floods have occurred with disastrous consequences. In June 2007, the cyclone Yemyin hit the Pakistani coast with torrential rainfalls affecting Gwadar and other places. On that occasion, a dam was flooded, killing people and leaving many others without their basic needs, such as housing and food [17]. Consequently, even though this kind of phenomenon is rare in the region, a thorough safety design is of the utmost importance.

6.4.5 Summary

Table 6.6 summarizes the main impacts.

6.5 Value conflicts and solutions

A variety of technical, social and other issues appeared to be involved in the construction and operation of Gwadar Port. These issues are about conflicting interests, goals and, ultimately, values. Responsible Innovation helps in designing, developing and managing innovative strategies involving all stakeholders in large technological projects, including port projects (see [2, 20, 21]). This requires an analysis and resolution of value conflict.

6.5.1 Value conflict analysis

In order to find value conflicts, one has to look first at the stakeholder values and the possible impacts that might affect these values. On the basis of the data in Tables 6.5 and 6.6 as well as Figures 6.5 and 6.6, the following value conflicts were found:

- The value conflict between the social development of Gwadar citizens and economic growth when it comes to the resettlement of people

Table 6.6: Possible impacts of Gwadar Port.

Possible impacts	Negative	Positive
Social impacts	Resettlement Loss of jobs (fishermen) Demographic changes	Gender inclusiveness
Environmental impacts	Mangrove forest destruction Livestock resettlement Air and noise pollution Seabed deterioration Oil spills	Agricultural development
Economic impacts	Competitors will lose the Chinese market (mainly Indian ports)	Economic development Increased employment Less shipping costs for exporters to China (EU)
Safety impacts	Terrorist attacks Earthquakes and tsunamis Floods	

- The value conflict between profitability and environmental friendliness when it comes to air and noise pollution
- The value conflicts between economic and political issues when it comes to safety

6.5.2 Value conflict solutions

The first two value conflicts have to do with the project itself, while the third one has international characteristics.

6.5.2.1 Solving internal value conflicts

The main reason for having conflicts between social development and economic growth and between profitability and environment is the fact that there is no fair power distribution in these policy games. The CNDRC holds the most power, while the citizens (CG) have much less. Consequently, we propose to improve the CG power base, giving them a stronger negotiation position. To this end, it is first and foremost necessary for the citizens to be organized in a strong institution representing all of them, allowing them to control or at least counterbalance the more powerful stakeholders. However, since the education level in the Gwadar area is low, it is very unlikely that people can do it themselves. Consequently, the Pakistani government should take up this challenge to help the citizens in creating and maintaining such an institution. Through this institution, the CG can ensure to be included in the decision-making processes of the Gwadar Port project and stand up for their values, while transparent negotiation may reduce value tensions.

6.5.2.2 Solving international value conflicts

There is a conflict between economic and political values concerning the United States, believing that the Gwadar Port is there to serve China's military desires in the area, and China, claiming that the port only serves its (and Pakistan's) economic interests. The solution we propose is to increase cooperation between the two countries under the auspices of a strong mediator, for example, the United Nations. However, due to the historical conflicts between the United States and China, this solution can be considered as a very unattainable target. Nonetheless, if this is achieved and the United States and China have the same view on Gwadar Port, other conflicts in other political contexts could also be tackled.

6.6 Improvements to the project

Large infrastructure projects such as ports, highways and airports are the engines of economic growth and development in every country. However, when such projects are designed with economic goals as the sole driving force, the ultimate economic goal may fail as a result of resistance from people or public opposition. Consequently, it is recommended to include other goals in the project considering different stakeholder interests and values, such as environmental or societal goals. That is where Responsible Innovation comes in, as it 'aims at designing, developing and managing innovation trajectories, including all stakeholders, analysing their goals and interests in terms of basic underlying values, and preventing and solving value conflicts between stakeholders' [2].

In the case of Gwadar Port, the main goals are strategic and economic, not social, while people are poor, social differences are great, access to education or health services is lacking and drinking water is scarce in the region [17]. Consequently, improvements in the project are crucial, including social goals and resolving the identified value conflicts. RI might help.

6.6.1 Stakeholder involvement

If a project like Gwadar Port has to include societal interests and values, the only way to do so is by directly involving the stakeholders. Project managers often confuse the consultation and participatory styles in this. RI aims to engage the stakeholders by making them participate in decision-making processes, rather than only asking them about their interests and views [22]. Therefore, in the Gwadar Port case, it is mandatory to include local communities in a participative way of decision-making to ensure that their basic needs in the areas of education, water, health and housing are covered.

6.6.2 Include sustainability goals

RI also ensures sustainability. As defined by the Brundtland Commission in 1987, 'sustainable development is the development that meets the needs of the present without compromising the ability of future generations to meet their own needs' [23]. In this sense, society and the environment are put on the agenda. As has been made clear, sustainability has not really been considered in the Gwadar project. This should be corrected.

6.6.3 Review ethical principles

In large infrastructure projects, ethical concepts should always be considered during design, construction and operation, which probably did not happen in the Gwadar project. According to Van de Poel and Royakkers [23], the main ethical principles to include are the following.

6.6.3.1 Availability of alternatives

In order to carry out RI, considering alternatives for the project is crucial. Even if people have not been involved, if they know that several alternatives have been scrutinized and that the selected one is the most appropriate, it is more likely that they will give their consent to the project and act in line with its objectives. No other alternatives serving the same goals were probably analysed in the Gwadar case.

6.6.3.2 Informed consent

If every stakeholder is fully informed about the potential risks of the project and they agree, we would meet the requirement of informed consent. The Gwadar Port project did not include all stakeholders, which is the reason that it can be concluded that this ethical principle was not followed. RI requires revising and amending the process in order to include stakeholders and ask for their consent regarding possible risks.

6.6.3.3 Precautionary principle

The precautionary principle claims that if there is (1) a threat, which is (2) uncertain, then (3) some kind of action (4) is mandatory [22]. In this context, the safety effects identified as a result of terrorist attacks, earthquakes and floods must be reviewed in order to find out what (additional) actions need to be taken to reduce these risks.

6.6.3.4 Risks and advantages are fairly distributed

If risks and advantages of a project are fairly distributed, RI would have been successfully implemented. Not so in the Gwadar Port case. Local people still lack basic needs, such as education, health services, a good sewage system and a reliable drinking water system, while the Chinese enjoy economic benefits from the port operation due to better access to the EU for their products. A fairer distribution of benefits and risks is imperative.

6.7 Implementation plan

All new requirements resulting from our previous discussions call for an adapted implementation plan, enabling the CNDRC to improve and proceed with the Gwadar Port project responsibly. The new plan consists of seven steps.

6.7.1 Step one

The first step is the clear and unambiguous identification and definition of the problem. It should also be determined for whom it is a problem and why it is problem.

6.7.2 Step two

In the second step, all actors involved should be identified and mapped. This step is crucial to the success of the project. An extensive stakeholder analysis will determine parties with a keen interest in the project (including the problem owner and the initiator of the project), but also actors that are affected by the decisions, even if they might not be interested in the project. In addition, such analysis will reveal which stakeholders have the power to influence decision-making and which parties are powerless.

6.7.3 Step three

The third step pertains to the identification of possible solutions and an analysis of involved technologies. The former refers to the range of options available and the options considered as good and viable, while the latter refers to the availability of suitable technologies and new developments.

6.7.4 Step four

The actual discussion or interaction with stakeholders to gauge different views and values is step four. In the participatory style, several methods can be used for this purpose, including structured meetings with stakeholders (or their representatives if, for instance, the local community is the actor), conferences, public opinion surveys, referenda, round table debates, public hearing/inquiring and so on [24]. An additional important aspect of this style is the creation of commissions that will represent specific groups, such as fishermen or farmers, and, by all means, the creation of a committee dealing with ethics and sustainability.

Special attention should be paid to groups of people that may lose more than desired, for instance, people who will be 'forced' to relocate. In this, the presence of a neutral mediator can be significant, for example in the form of a multi-stakeholder advisory committee (MAC) as Ravesteijn et al. [2]) proposed. The independent mediator would facilitate decision-making since its target would be to address disputes and achieve acceptable agreements. During this work, the basic rules of decision-making must be established in order to achieve fair treatment for all participants during each round [9].

This step is of the utmost importance to the successful implementation of the plan. Policies that include the people are significantly more likely to gain public approval and support during the phases of design, implementation and operation.

6.7.5 Step five

Responsible Innovation and sustainability require attention to fundamental values in relation to impact. The values can be found in step four, while step five concerns the impact, which requires environmental, social, economic and risk impact assessment studies. If we have in-depth knowledge of the potential effects of planned infrastructure construction, the design and implementation will be improved, while possible obstacles can be addressed, which contributes to the success of a responsible and sustainable project.

In order to cultivate a participatory attitude of stakeholders, it is necessary that the entire procedure is accessible to the public and free of all suspicion. This implies transparent procedures and disclosure of relevant information. For example, directly and less directly involved parties should be given the opportunity to consult the results of the impact assessment studies, which should be available online or generally easily accessible.

6.7.6 Step six

The sixth step is almost the same as the fourth step (participation and discussion). The difference lies in the consideration and resolution of the value conflicts in this step, based on the new information provided by the impact assessment studies.

6.7.7 Step seven

The seventh and final step concerns the implementation of the project in collaboration with the stakeholders. Continuous monitoring in all phases of the project (design, implementation and operation) needs special attention, enabling the identification of all kinds of emerging problems that must be incorporated in decision-making. These types of projects always have an iterative character, for example, in this step, the appearance of new stakeholders with potentially different requirements, objectives and ideas or the alternation of position (power, interest) in comparison with existing stakeholders [8].

6.8 Conclusions

Policy- and decision-makers of large infrastructure projects are frequently focused on the potential economic benefits that the implementation of the construction will induce. Quite often, other relevant important aspects are not taken into consideration and, consequently, the design is usually incomplete. The absence of impact assessment studies revealing possible effects, both positive and negative, is a noteworthy example. Responsible Innovation offers a comprehensive approach. It covers the acquisition of information of a wide range of possible effects and the evaluation of these effects in terms of moral values, such as equity, privacy, well-being, sustainability, accountability and so on. Subsequently, it entails the utilization of the new knowledge as a functional requirement for the design and development of a project.

The present chapter analysed the construction of Gwadar Port in Pakistan from the perspective of the CNDRC. Although this project has not been set up as such, Responsible Innovation has been used to analyse it and suggest improvements.

Project complexities relating to technology, organization and the social context have been identified and discussed, revealing reasons as to why the project might not have proceeded according to the initial plan, especially because not all stakeholders were included from the start. Impacts of the project were determined, mainly in relation to economic, social, environmental and safety effects of the construction. An extensive stakeholder analysis has been carried out, listing all parties and their interests and values, and enabling the identification of value conflicts, which are, ultimately, a major threat to the success of the project.

Based on this analysis, improvements to the Gwadar Port project in Pakistan were formulated. A seven-step implementation plan for the project, including improvements, has been proposed, which should serve as a guideline for the CNDRC and steer the construction of Gwadar Port in such a way that profit and prosperity are achieved without jeopardizing the natural environment.

Acknowledgements

This chapter is based on a paper that the authors successfully concluded the Management of Large Technological Projects course, which was part of the joint Engineering and Policy Engineering master and exchange programme of the Harbin Institute of Technology and Delft University of Technology. Wim Ravesteijn, the module manager of this course, has thoroughly edited the text.

References

[1] Van den Hoven, J., Wildschut, E. & Bakker, E., Responsible innovation at port cities. *Boll. del Dipartimento Conserv. dei Beni Archit. Ambient.*, 12(1), pp. 765–770, 2012.

[2] Ravesteijn, W., He, J., & Chen, C., Responsible innovation and stakeholder management in infrastructures: The Nansha Port Railway Project. *Ocean & Coastal Management*, 100, pp. 1–9, 2014.

[3] Gwadar Port Authority, http://gwadarport.gov.pk/. Accessed on: 8 December 2015.

[4] Chen, Z., The authority of Gwadar Port will be shifted to China. Xilu, 4 September 2015, http://www.xilu.com/20150804/1000010000855090.html. Accessed on: 8 December 2015.

[5] Kazim H., The Karakoram Highway: China's Asphalt Powerplay in Pakistan. *Der Spiegel International*, 17 July 2012, http://www.spiegel.de/international/world/china-expands-karakoram-highway-to-pakistan-a-844282.html. Accessed on: 8 December 2015.

[6] Pasquali V., The poorest countries in the world. *The Global Finance Magazine*, 29 November 2015, https://www.gfmag.com/global-data/economic-data/the-poorest-countries-in-the-world?page=12. Accessed on: 8 December 2015.

[7] Baloch A., Gwadar's importance. *Daily Times*, 24 October 2015, http://www.dailytimes.com.pk/letters/24-Oct-2015/gwadar-s-importance. Accessed on: 8 December 2015.

[8] Bosch-Rekveldt, M., *Managing Project Complexity: A Study into Adapting Early Project Phases to Improve Project Performance in Large Engineering Projects*. Delft: Delft Centre for Project Management, 2011.

[9] Enserink, B., Hermans, L., Kwakkel, J., Thissen, W., Koppenjan, J. & Bots, P., *Policy Analysis of Multi-Actor Systems*. Lemma: The Hague, 2010.

[10] Bruijn, H. de & Heuvelhof, E. ten, *Management in Networks: On Multi-Actor Decision Making*. Routledge: London, 2008.

[11] Jiang, Y., Behind the 'One Belt, One Road' Policy. *The Southern Urban Daily*, 10 November 2014, http://news.southcn.com/china/content/2014-11/10/content_111863207.htm. Accessed on: 8 December 2015.

[12] Wang, Q., The stories of Gwadar Port and China, 21ccom, 13 September 2015, http://www.21ccom.net/articles/world/qqgc/20150813127872.html. Accessed on: 8 December 2015.

[13] Zheng, Q., The Gwadar Port is about to be put into operation, *21st Century Economy*, 18 April 2015, http://news.sohu.com/20150418/n411458156.shtml. Accessed on: 8 December 2015.

[14] Lei, T. van der Lei, T. & Ligtvoet, A., Value focused thinking: An approach to structure company values for asset risk management. *Proceedings of 7th World Congress on Engineering Asset Management*, pp. 8–10, Daejeon, Korea, 2012.

[15] PAK-EPA, http://environment.gov.pk/. Accessed on: 8 December 2015.

[16] Yousaf, T.A., Is Gwadar Port an Economic Haven for Balochistan and Pakistan? June, pp. 1–59, 2012. https://lup.lub.lu.se/luur/download?fileOId=3224623&func=downloadFile&recordOId=3224614. Accessed on: 21 May 2021.

[17] Government of Balochistan and IUCN Pakistan, Gwadar: Integrated Development Vision, IUCN Pakistan, Quetta, Pakistan. viii+70 2007.

[18] Owen B., Pakistan: A failed state or a clever gambler? BBC, 7 May 2011), http://www.bbc.com/news/world-south-asia-13318673. Accessed on: 8 December 2015.

[19] Sial, S., The China-Pakistan Economic Corridor: An assessment of potential threats and constraints, 2014. http://pakistanhouse.net/wp-content/uploads/2016/11/cpec.pdf. Accessed on: 21 May 2021.

[20] Ravesteijn W., Responsible Innovation and Port Development with Reference to the Rotterdam Port Extension 'Maasvlakte 2' Project, working paper, 2014.

[21] Grigalunas, T.A., Luo, M. & Chang, Y., Comprehensive Framework for Sustainable Container Port Development for the United States East Coast, Year One Final Report, URI-TC Project, No. Grant 536106, 2001.

[22] The Interorganizational Committee on Guidelines and Principles for Social Impact Assessment, Guidelines and principles for social impact assessment. *Environmental Impact Assessment Review*, 15(1), pp. 11–43, 1995.

[23] Van de Poel, I. & Royakkers, L., *Ethics, Technology, and Engineering*. Oxford: Wiley-Blackwell, 2011.

[24] Rowe, G., & Frewer, L.J., Public participation methods: A framework for evaluation. *Science, Technology & Human Values*, 25(1), pp. 3–29, 2000.

CHAPTER 7

Do the Chinese bring Sustainable Development to Africa? The Silk Road and the Port of Bagamoyo in Tanzania

Andreas van Giezen, Larissa Eggers, Alexander Peters & Rahul Prasad

Delft University of Technology, The Netherlands

Abstract

This chapter will discuss the Port of Bagamoyo and the Special Economic Zone project in light of the One Belt, One Road initiative, Chinese activities in Africa, and Tanzania's as well as Africa's recent economic growth. The focus will be on sustainability in all dimensions; the main question is whether – and to what degree – the project can be considered sustainable in social, environmental and economic terms. It will conclude with recommendations as to how the project could be improved for the benefit of the people of Bagamoyo and beyond.

Keywords: Port of Bagamoyo, the Chinese in Africa, Sustainable Development

7.1 Introduction

The One Belt, One Road (OBOR) initiative taken by China is a massive development plan consisting of a multitude of projects along the historic overland Silk Road as well as the new 21st Century Maritime Silk Road from China through South and Southeast Asia and East Africa to Europe. The main goals of this initiative are outlined in the vision plan released by China in 2015; it aims to 'synchronise development plans and [to promote]...joint actions amongst countries along the Maritime Silk Road...to build a peaceful and prosperous 21st Century Maritime Silk Road' [1]. Within this initiative, ports in Asia and Africa will be developed into trade and industry hubs.

One such project is the development of the Port of Bagamoyo (PoB), Tanzania, on Africa's eastern coast. A Chinese and Omani consortium is currently building an almost entirely new deep-seaport with twice the capacity of Tanzania's current main port of Dar-es-Salaam, some 80 kilometres further south. The idea behind the investment in the former slave harbour is to create a new port as well as a so-called Special Economic Zone (SEZ), which ultimately is to become the leading shipping and logistics centre of East Africa. Hopes are high that this project will be a motor for development in Tanzania, its effects extending way beyond the town of Bagamoyo. The project is a major part of Tanzania's development strategy, and the government emphasizes their admiration for China as a role model for economic development [2]. However, the start of the project was delayed several times, and the Tanzanian government experienced setbacks. The most prominent setback was the surrender of its ownership of the port due to a failure to raise funds for compensation to the current inhabitants as a result of their displacement from the new port's location [3]. According to recent news reports, Tanzania sought renegotiations with the investors in an attempt to reclaim partial ownership [4], which would allow them to ensure higher benefits for the local population.

Combined with past accounts of human rights violations, business malpractices and a massive influx of Chinese labour along with Chinese investment in Africa, concerns were raised about the prospect of success of the PoB and SEZ project. It can be questioned whether the investments in the PoB will lead to Sustainable Development, i.e. economic growth in the region combined with economic and social integration and protection of the environment.

This chapter will discuss the PoB and SEZ project in light of the OBOR initiative, Chinese activities in Africa and Tanzania's (also, Africa's) recent economic growth. The focus will be on sustainability in all dimensions, the main question is whether – and to what degree – the project can be considered sustainable in social, environmental and economic terms. It will conclude with recommendations as to how the project could be improved for the benefit of the people of Bagamoyo and beyond.

In order to answer this research question, it is vital to analyse what the OBOR initiative exactly entails and how the Chinese government plans to apply it. Furthermore, we look at the literature to familiarize ourselves with the commonly accepted definitions of sustainability, development and responsibility, which will constitute a conceptual and theoretical framework for analysis and evaluation. Since this chapter contains a case study, we will specifically investigate the PoB and SEZ project. The scarcity of up-to-date data on the project has limited our study to a macro level; for this reason, the reputation and perceptions of China in Africa will be investigated based on all of Africa rather than Tanzania specifically. Our analysis will consist of an inventory of all stakeholders involved and the impact each of them has or will experience, as well as a discussion on the development effects of the project.

The OBOR initiative will be described in Section 7.2. Conceptual and theoretical notions will be elaborated upon in Section 7.3. The case description follows in Section 7.4. China's reputation and perceptions in Africa is the subject of Section 7.5. Section 7.6 gives our analysis, followed by the conclusion with recommendations in Section 7.7. Section 7.8 discusses the limitations of our case study and concludes the chapter.

7.2 China's One Belt, One Road initiative

In the autumn of 2013, China's President and CCP General Secretary Xi Jinping announced a long-term strategic framework of building a 'Silk Road Economic Belt' and a '21st Century Maritime Silk Road.' These frameworks are referred to as the 'One Belt, One Road' or the OBOR initiative, and the aim is to promote peaceful cooperation and development around the world [5]. Xi Jinping announced that the OBOR initiative could potentially benefit 4.4 billion people or roughly 63% of the world's population [6].

Since 2016, more than 100 countries announced their support and willingness to participate in OBOR projects [7]. Besides this, multiple international organizations such as the United Nations (UN) and Asia-Pacific Economic Cooperation (APEC) have incorporated and reflected on the OBOR initiatives in their resolutions and documents [5]. The OBOR initiatives are seen as a modern representation of the ancient Silk Road. This ancient Silk Road connected the East and West more than 2000 years ago and enabled goods and knowledge to be exchanged along it [8]. In this modern time, the ancient Silk Road is particularly well known due to the stories of the explorer and merchant Marco Polo, who travelled along the road around the end of the 13th century.

State documents released by the People's Republic of China (PRC) show that the OBOR initiatives aim at improving the trade and connectivity within Asia as well as between Asia and the continents of Africa and Europe. Furthermore, the OBOR initiatives are claimed to focus on (regional) Sustainable Development, improving the integration of economies and their nearby states, and creating stability and peace in regions around it [5, 9].

China intends to do this by constructing and redeveloping ports, roads and pipelines within China itself, Asia and Africa [5]. To accomplish this, China originally established a US$40 billion fund to finance OBOR-related projects. An extra US$20 billion became available in 2015 and it is expected that around US$100 billion will be made available in the coming years [6].

State-councillor Dai Bingguo said that OBOR projects will adhere to the paths of peaceful development and can only become successful if there is 'mutual trust, equality, mutual learning, and a win-win cooperation between the different states' [6]. According to Johnson [6], these norms and guidelines can be traced back to China's long cultural history. There are two points where China's foreign policy approach and the OBOR initiative come together. The first one is that China's further development can only be achieved by peaceful means and win–win situations. This means that if China's neighbouring countries or other trading partners do not further develop, China's development could stall as well. The second point is related to milestones that China aims to achieve by 2021 when the Chinese Communist Party (CCP) celebrates its 100th anniversary and 2049 when the PRC celebrates its 100th birthday [6]. The milestones to be achieved by 2049 state that China must be a 'socialist, strong, prosperous, democratic, and culturally advanced harmonious country.'

Concluding, the OBOR goals can be summarized as follows. On a macroscopic level, the initiative is a global-scale initiative that ensures China's economic partnerships with numerous countries while aiming to improve trade and connectivity within Asia, as well as between Asia and Africa and Europe. This is mainly carried out through investment in large infrastructure projects such as ports, roads, railways and pipelines. Peaceful cooperation, trust and sustainability are emphasized to be the backbone of these economic development plans. Potentially, the initiative will benefit about half of the world's population. These macroscopic goals are further specified for the micro level. Cultural exchange and mutual learning, understanding for other cultures and respect for local environments are also mentioned as goals.

7.3 Sustainable Development and ports

'Humanity has the ability to make development sustainable – to ensure that it meets the needs of the present without compromising the ability of the future generations to meet their own needs' [10].

Sustainable Development came onto the global agenda with the 1987 report 'Our Common Future' by the World Commission on Environment and Development (WECD) of the UN, also known as the Brundtland Report. The above definition emphasizes intergenerational justice, i.e. our economic activities today should not destroy the source of life for future generations. The report continues with a note on intragenerational justice – 'sustainable development requires meeting basic needs of all and extending to all opportunities to fulfil their aspirations for a better life' – emphasizing that poverty is not only 'an evil in itself,' but makes the world vulnerable to ecological catastrophes, which endangers the goal of preserving a liveable planet. Thus, Sustainable Development requires world inequality to be reduced through economic growth and increasing living standards for the world's poorer countries.

Sustainability is typically discussed along the dimensions of people, planet and profit; or social, environmental and economic issues. Development itself is closely linked to economic growth, often being defined as the transition from a state of stagnation to ongoing growth. A global framework for Sustainable Development is given by the UN's Sustainable Development Goals [11]. This set of 17 goals spans a large array of topics and applies to all countries in the world, unlike its predecessor, the Millennium Development Goals, which aimed at developing countries, to which Tanzania belongs.

Looking at a port project, a combination of two goals can generally be identified. Ravesteijn, He and Chen [12] emphasize the dual character of the goals of port projects. On one hand, a port project has an entrepreneurial goal of providing good services to their customers; on the other, they aim to increase social welfare for the citizens of the port area. Dwarakish and Salim [13] also point out the importance of ports as development motors; it is through the provision of jobs that they contribute to economic and social development. They contribute to the integration of their location into the world market, opening up opportunities for trade with natural resources as well as manufactured goods. Through income for local residents, education for high-skilled jobs, social development and integration can be supported. This does not happen inevitably though, it needs to be shaped by the government. Additionally, the effects of port construction and operations on the environment can be detrimental. Not only incidents like oil leakages but day-to-day operations of a port have negative consequences on the surrounding ecosystem.

7.4 Tanzania and the Port of Bagamoyo

Albeit much debate pertaining to its genesis, Bagamoyo or *Bwaga-moyo* (in Swahili), which means 'lay down your heart,' is either related to the slave trade that passed through the region – where the slaves were told to 'lay down their hearts' at Bagamoyo before being sent overseas only to never see their homes again – or to the oasis where, in the early 19th century, 'laying down their melancholy,' the porters rested after carrying loads on their shoulders brought to the town from the interior to trade [14, 15]. In either case, despite there being only little evidence to substantiate the claim that the town was a major slave trading centre, Bagamoyo served as a major trading port [16] and two centuries later, the Chinese, as part of their OBOR project, are developing the PoB.

This section zooms in on the situation in Tanzania and Bagamoyo regarding the current state of the economy and the environment, as well as plans that were made regarding the PoB and SEZ; social issues are addressed in a tentative stakeholder analysis.

7.4.1 Tanzania: Economic and political situation

Many African countries have seen substantial economic growth in the last one or two decades. Tanzania is among them and has recorded economic growth of 6.9% in 2016 [17], which makes it the second-fastest growing economy in Africa after the Ivory Coast. Part of this growth can be attributed to Chinese trade and investment that also grew substantially. In addition to an upward economic trend, relative political stability is an asset. While many other African countries have been through various forms of political and regional instability (such as the genocide in Rwanda, the continuing conflicts in Somalia spreading into Kenya, etc.), Tanzania has been a peaceful and stable country within the region [18].

Around 80% of the labour force is active in agriculture, and with education among youth in rural areas being comparatively low, there are limited skilled workers for the (high-tech) operating firms in Tanzania [19]. A case study into the finances of the country, capital inflows and economic growth in Tanzania shows that 'financial development in Tanzania follow[ed] [economic] growth' [20].

7.4.2 Bagamoyo's people and environment

Bagamoyo lies in an environmentally rich land with the Ruvu River to the north, the Lazy Lagoon Island to its south and with mangrove forests on either side. The people living in the Bagamoyo area (approximately 300,000) utilize this rich environment and make a livelihood out of agriculture, fishing, mariculture (or seaweed farming) and prawn farming, and, along with trade and tourism, they form the main economic activities. Nevertheless, poverty is still rampant in Bagamoyo, which makes the local communities vulnerable and easy prey for exploitation; for instance, even though tourism has created new job opportunities and businesses, its development has led to exploitation of resources and income inequality [21]. The area's mangrove forests – an environmental asset to the area and an essential part of the mariculture – are under threat by the construction of a port, first, for the large space it can occupy, possibly engulfing the mangroves, and second, by the pollutants that almost inevitably will be released into the water in a deep-seaport.

7.4.3 The Port of Bagamoyo and the SEZ project

The Port of Bagamoyo is located less than 80 km north of Tanzania's major port town Dar-es-Salaam and about 300 km to the south of Kenya's Port of Mombasa. In a study that explored the demand for port capacity in East Africa, considering Bagamoyo as an alternative location of a new port, two conclusions are prominent: the port in Dar-es-Salaam could only function as a regional lower-tier hub distribution centre for the trade of Tanzania and its hinterland; and second, compared with Dar-es-Salaam, the transportation cost by road for a tonne of transit cargo is US$0.72 cheaper for Bagamoyo. The PoB as a regional hub would serve as the gateway to the inland and therefore, a gateway to Africa for travellers and traders on the Silk Route [22]. The potential of Bagamoyo as a port is thus not only due to its proximity to existing and planned infrastructure to the inland but also due to its location within the whole region (Figures 7.1 and 7.2).

The PoB and SEZ project's development began as a three-way partnership between China Merchants Port Holdings (CMPH), Oman's General State

Figure 7.1: Location of Bagamoyo. (Source: Google Earth).

Reserve Fund (GSRF) and the Tanzanian government for which the framework agreement was signed on 24 March 2013. However, ‘financial constraints [...] have forced Tanzania to miss out on ownership of the US$10 billion Bagamoyo Port and Special Economic Zone project’ [3]. In exchange for an undisclosed shareholding in the project, the Tanzanian government was to raise US$28 million in order to compensate the 2180 registered landowners (who would be displaced from their land that the project would be founded on); however, the government was only able to raise US$1.5 million. Tanzania is thus at risk of losing ownership and influence in the project – while the CMPH stepped in to fill the void and found the money needed for landowner compensation – and will only (or at least, is purported to) benefit from taxes on the land and occupancy by investors. In a statement, managing director (of CMPH) Hu Jianhua said that the company would run Bagamoyo as one of its overseas ports [3].

The Tanzanian–Chinese relationship has been developing for some time, and the Chinese investments in Tanzania have already grown to US$6.62 billion. Investments include projects in agriculture, tourism and ICT with 150,000 jobs being generated and over 350,000 individual actors already being engaged in trade activities between these countries [23]. The intention of China is to increase the investments even further with the PoB and SEZ project, and according to *The Oriental Review*, China has loaned Tanzania US$7.6 billion for a central corridor railway, for which the contract has been awarded, not surprisingly, to a Chinese contractor but for US$9 billion [18]. The PoB,

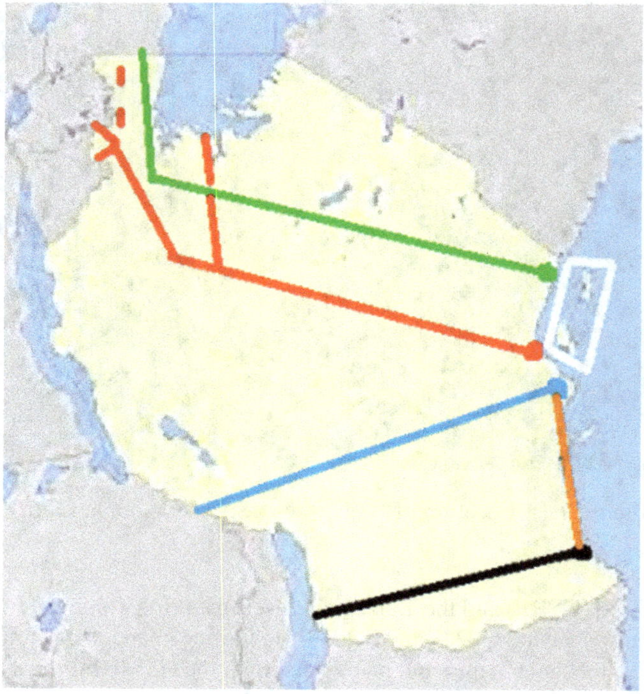

Figure 7.2: Infrastructure links to be upgraded (or developed) in Tanzania. (Source: Korybko, 2017.)

connecting to the central corridor (red line in Figure 7.2 and the port being the red dot), will be located 75 km north of the Port of Dar-es-Salaam and 10 km from the town of Bagamoyo. The port will be designed to handle 20 million TEUs per year. The central corridor will be developed to connect the port to the landlocked countries in the hinterland. Attached to the port is the SEZ where some 190 industries have been marked for development, which is anticipated to grow to 700 industries when the port is fully developed [3]. The vision for the SEZ (by CMPH) is for a US$120 million industrial park, a US$70 million tourist park, free port facilities worth US$90 million, a free trade zone worth US$70 million, a science and technology park worth US$50 million, an international business centre worth US$70 million and an industrial building worth US$20 million [23].

Besides the aforementioned US$9 billion central corridor, other major routes or infrastructural links are being developed that are either 'Chinese funded or Chinese contracted in one way or another' [18]. Korybko gives a comprehensive coverage of these structures: the Uganda–Tanzania oil pipeline (green line) and Tanga Port is the green dot; Tazara line (blue line) and Dar-es-Salaam is the blue dot; connecting Tanzania and Zambia is the Mtwara development corridor (black line) and Mtware is the black dot; and the Dar-es-Salaam–Mtwara oil pipeline is the orange line. These links can ensure connectivity between the landlocked countries and the major ports, including the PoB, facilitating means of transportation of resources and products for exports and imports. It can be argued that the southern route from Mtwara will enable access to iron ores and the coal mines in Malawi, which at the same time will aid the Russians in transporting uranium from the Mkuju River. The other southern line will be connecting Dar-es-Salaam with the copper-rich country of Zambia. The northernmost line will tap into the oil reserves of Uganda and the central corridor will connect Rwanda, Burundi and the DR Congo – a prosperous country with regard to exploitable resources and labour force – to the PoB. It can be discerned, therefore, that with the central corridor, all three developing countries with huge potential inexpensive labour markets and resources are provided an access to the free trading market and for China to the OBOR.

Although the SEZ in Bagamoyo is being built close to an endangered mangrove forest, there was no mention regarding environmental impact reduction in the plans for the construction of the SEZ in Bagamoyo [24, 25].

7.5 China's influence on Africa

This section serves the purpose of investigating China's economic relationship with Tanzania and other African countries, as well as the opinion(s) of the local populations. While we are – of course – aware that Africa is a large and diverse continent, the similarities between (sub-Saharan) African countries and the way they are seen from abroad, as well as the availability of data sources concerning several countries jointly, caused the decision to sometimes look at 'Africa' as a whole rather than just Tanzania.

7.5.1 History and background of China's investments in Africa

Projects like the PoB and the surrounding SEZ (or the whole OBOR initiative) need to be seen as state-funded commercial activities, not development aid (ODA: Official Development Aid). Dreher et al. [26] argue that much of the Western criticism towards China's actions in Africa is caused by the misunderstanding of these cash flows as aid. This evokes the false expectation of altruism while the motivation behind such investments is clearly business-driven. However, not qualifying as development aid does not imply inferiority; authors like William Easterly [27] and Zambian-born and raised Dambisa Moyo [28] even argue that investment and infrastructure-driven development as practised by China are superior and even the best way to help sub-Saharan Africa out of poverty.

China has always emphasized the mutuality of their financial activities in other countries, as already laid out in 1964 by the PRC's first Premier Zhou Enlai in his '8 Principles of Chinese Aid' as quoted by Bräutigam [29] and continued in the 2015 kick-off document for the OBOR initiative [1].

The framework to China's (previous) investments abroad is given by the so-called 'Beijing Consensus' – China's approach to economic development, which is first and foremost applied at home, serves as a development model for other countries, who either adopt it by themselves or in collaboration with China. Both are the case in Tanzania, where China is cited as a general role model for development and is, as we know, involved in specific projects like the PoB and SEZ and others [2]. The term 'Beijing Consensus' was coined by J.C. Ramo in 2004 in reference to the Washington Consensus, the American development approach [30]. The latter is a neoliberal approach including fiscal discipline, free markets, privatization and deregulation, but also democracy and political freedom [31]. This approach, often attempted to be imposed top-down, has failed in many developing countries. The Beijing Consensus on the other hand has been met with much more sympathy there. It is described as 'economic growth without political liberation' [32], which could explain the preference for this approach held by many political leaders in non-Western countries. Economically, it is a much more incremental approach than the top-down Washington Consensus, based on innovations and trial-and-error. Furthermore, it rejects GDP as the sole indicator for growth, but also includes people-related and sustainability measures [32]. Arif Dirlik, a history professor with a focus on China at Oregon University, argues that the Beijing Consensus could provide a common global framework for economic relationships that is able to recognize political and cultural differences [33].

7.5.2 Perception and criticism of China's actions as reported by locals

Chinese investments over the last decades in healthcare, education, infrastructure and even military aid, despite the non-intervention model, have brought opportunities for local development to Africa [34, 35]. Yet, studies demonstrate the local citizens in the receiving countries are not universally appreciative of China's involvement on the continent [34, 36]. With now over a million Chinese citizens living in Africa, the rising tensions between the local populace and the new Chinese immigrants are a cause for concern. The attitude of the Africans towards China is summarized by one study: 'some Africans have pondered if China has become a development partner, an economic competitor or simply a new hegemony to Africa' [35].

Extensive surveys indicate that attitude is similarly conflicted, extending even to the elites running the countries [34, 36]. Openly welcoming the Chinese officials for the 'free' money they represent, active policymaking tends to disfavour Chinese involvement [35, 37]. In Zimbabwe, landing fees for Chinese citizens are nearly double those of European countries, higher even than for citizens from the UK and Canada, countries Zimbabwe is hostile towards [37]. In Ghana, policies towards illegal mining in the gold industry specifically target Chinese people even though they are not the only ones there [38]. Interestingly, the development money of the Chinese in these countries tends to surpass that of the European countries [35].

The situation on a local level tends to be similar with segregation and racism as main concerns. In Antananarivo, Madagascar, of the 60 Chinese-owned businesses, none hire African workers, and only a few serve African clients [35]. The latest influx of Chinese people tends to view the Malagasy as 'children with a poor work ethic and a lack of reliability,' and most immigrants have wanted to return to China in three to five years despite enjoying living in Madagascar [39]. On the other hand, anti-Chinese sentiment is rising among the Malagasy, which has led to multiple riots. The situation in Madagascar is worrying and is representative of many other African countries [35, 39].

Causes for African resentment are Chinese business practices, often with China's backing. Examples are numerous and often involve a general policy of recklessness and financially unsound behaviour – as losses are covered by the Chinese government – effectively eliminating African competitors [35]. Additionally, skilled low-cost workers are flown in from China, partially government funded, over training local African workers [37]. For many Africans, this goes to show that China is only interested in China and hardly cares for the local populace, which is enhanced by Chinese disregard for working conditions and human rights [35, 39].

Chinese business ventures in Africa are depicted as neo-colonialist, a situation in which resources are extracted and shipped to China and the manufactured goods shipped back to Africa, resembling colonial-era practices of the Europeans [35, 37]. These practices raise concerns by many Africans, both due to the goods often being regarded as low quality as well as the limited opportunities for African countries to use their own resources within this model [36, 39].

Another complaint revolves around environmental concerns mainly voiced in Western countries. Aside from direct environmental risks, like the damage to the mangrove forest in the case of the SEZ of Bagamoyo, ports themselves account for a large ecological footprint, for example, the Port of Rotterdam accounts for nearly 20% of Dutch CO_2 emissions [40, 41]. Whereas many Western ports have implemented policies to reduce their impacts, often encouraged by local governments, this is not a common practice in Africa, with the SEZ of Bagamoyo as an example. Albeit China's own policies regarding the environmental impact of ports are not as numerous or refined as Western policies, it is to be hoped that China implements them in their overseas construction plans within the OBOR initiative [24, 42, 43].

The last complaint often voiced revolves around the corrupt practices of the Chinese. Although these accusations are hard to prove, Chinese involvement tends to mitigate the positive effects on perceived corruption that World Bank projects have in an area [37, 44]. Concerns about undue political influence and backroom deals are prevalent although these are not uncommon in countries where immigrants exert influence in local government [39]. Other accusations are more serious, including criminal activities such as illegal ivory smuggling by Chinese officials [35, 40]. Even though many of these grievances are unsourced, it shows the growing resentments among Africans. Chinese government officials have admitted that their grasp on ethnic Chinese workers is rather limited and have expressed concern for the situation [35].

Still, views held by the Tanzanian population appear positive. In a 2014 study in which people were asked if their views towards China were generally favourable or unfavourable, Tanzanians showed the highest approval rate (77%) and the lowest disapproval rate (10%) of all eight African countries included in the survey. Furthermore, 58% of Tanzanians agreed that China will replace or has replaced the United States as the world's leading superpower [45]. According to another study, between 71% and 80% of Tanzanians have a favourable view of China, while 71% agree that China has a large influence on their country's economy. The positive views are attributed to China's investments in infrastructure and development by roughly one-third of the respondents [34].

7.6 Developmental effects and expectations

Returning to the question of whether the PoB and SEZ will contribute to Sustainable Development in Bagamoyo and its surroundings, several points will be discussed in this section. As not much information about the project itself could be found and the project is still in its early stages, it is difficult to give a comprehensive account. We will start with an analysis of the impact of the project for all relevant groups followed by some general considerations.

7.6.1 Interested parties and impact

The actors and stakeholders concerned with the PoB and SEZ project are diverse as appears from the brief account that follows. It shows the key players and the people profiting or suffering from the initiative.

1. China Merchants Port Holdings (CMPH). The main investor in the PoB and SEZ project and will run the port as one of its overseas ports. Management of the port will be executed by CMPH without any

input from the Tanzanian government or port authorities. The Port of Bagamoyo will be part of the OBOR project from China.

2. Oman's General State Reserve Fund. The other partner of the PoB and SEZ project and will reap the benefit according to an undisclosed agreement between the partners; no indication of interest other than generating revenues, the obvious interest of an investment fund, was found.
3. The Government of Tanzania. No longer a partner in the PoB and SEZ project and as mentioned before will only benefit from secondary effects like taxes on the land and occupancy by investors and increased economic activity. However, the government's interest in the care for the Tanzanian populace and the Sustainable Development of Tanzania persists as well as their admiration for China.
4. Registered landowners in the port area (including homes and farmland). All 2180 landowners will be compensated for their losses, a total amount of US$28 million (on average approximately US$12,850 per landowner) has been raised. However, there are many examples from history and around the world where the compensation and relocation of people have resulted in many difficulties and equity issues.
5. The local community of Bagamoyo. Comprises mainly farmers and fishermen and also people benefiting from developing tourism in the area. Their lives will possibly change substantially through a port construction in their proximity, on the one hand, through destruction of traditional livelihood and by the changes in the environment due to the PoB and SEZ project and its subsequent operations. On the other hand, the proximity to the port could bring new jobs to the community. Additionally, the port itself will also attract an influx of people – workers – to the area and may provide the local community with new business opportunities.
6. The Port of Dar-es-Salaam. Approximately 80 km south of Bagamoyo; currently Tanzania's main port but significantly smaller than the planned Port of Bagamoyo. However, Chinese investors have plans for Dar-es-Salaam as well – 'As for Dar-es-Salaam, Chinese businessmen from Jiangsu Province (home of the Eurasian Silk Road terminal, Port of Lianyungang) committed to investing US$5 billion in the region over the next five years' [18]. The vision is to develop the Port of Dar-es-Salaam and the link to the southern landlocked countries (including Zambia). It is therefore not expected that Dar-es-Salaam will be downgraded with the construction of the Port of Bagamoyo.
7. Hinterland (landlocked countries: Uganda, Rwanda, Burundi, DR Congo, and Zambia). These countries are currently without any facilities to bring their products to the international market in an efficient way. With construction of the infrastructure links and the PoB and SEZ

project, these countries will most likely have the means to bring their products and resources to the free market. However, like the Russian uranium mine along the Mkuju River, the companies exploring and possibly exploiting the natural resources of these countries are often foreign.

7.6.2 A Sustainable Development motor?

Within the various OBOR projects, China states it will support cultural exchanges and cooperation on all levels and encourages people from different cultures to learn from each other, blend with each other, understand each other and appreciate qualities in other cultures [5]. The vision document puts much emphasis on 'mutual benefits' and 'joint action.' On the other hand, the vision document also explicitly states that 'Chinese enterprises will be guided to participate in the construction and operation of ports' [1]. Combined with the previously mentioned reports of Tanzania losing ownership of the project, this raises concerns that the project might make Tanzania more dependent on Chinese support and goodwill.

Still, China's approach to development seems to be promising in Africa. It contrasts with the often-criticized Western approach, which is always seen with reference to the colonial past and has a reputation of being paternalistic with top-down decisions and tied to requirements such as democratization. The Chinese, on the other hand, purport to effectuate a different approach, both in the Beijing Consensus as well as the OBOR vision. They look for mutual benefits and cooperation; the vision document implies that two actors (China and the country receiving the investment) should look at each other as equals. 'Mutual benefit' is often criticized from the side of a nation who would misinterpret investments as aid; for them, 'mutual benefit' would mean that the motive of the other party is not altruistic. Since total benefits are a function of the synergies between the extent of independent benefits, it can be argued that the Chinese are taking things more seriously – truly striving for the success of their projects and establishing feedback and accountability, the lack of which economist William Easterly identifies to be the biggest reason for the failure of development aid. Presumably, economically Sustainable Development could truly be the goal of the Chinese, as they are interested in sustained economic relations. Social and environmental sustainability are more difficult points.

Based on the recently experienced economic growth in Tanzania, the ability of the Tanzanians to increase their capacity to produce goods and services and hence creating economic growth resulted in further financial development. Financial development can improve the positions of financial intermediation in the economy and create further investment, thereby inducing aggregate demand growth and more production, and so on, in the subsequent rounds of the circular flow. Also, with the large Chinese investment in the country, it can be argued that there will be further aggregate demand growth (and therefore economic growth) via the Keynesian multiplier process. However, the increase in aggregate demand will come into motion provided the local population enjoys employment and wage growth sufficient to create demand; and hence will be subject to the condition that in the new industries marked to function in the SEZ and the industrial parks, Tanzanian locals are given employment. However, as discussed in Section 7.5, the business model of Chinese (state-owned) enterprises in Africa is often neo-colonialist and value-extracting in nature. Besides, Chinese businesses in African countries have often resulted in side-effects of exploitation of the local community by smuggling ivory and gold. Therefore, by turning Tanzania into a transportation facilitator, without giving opportunities to Tanzanians, China would only be exploiting Tanzania. However, it will also be up to the Tanzanians themselves to take up the challenges and opportunities the new infrastructure projects may bring along.

As mentioned before, industries have marked their spot in the SEZ to open for business. SEZ was the strategy that put China on the world trade and industry map, which helped them to 'industrialise by attracting foreign investment[s].' In 2006, China decided to take its special economic zone overseas including Zambia, Ethiopia, Mauritius, Nigeria and Egypt. Although there is a lack of comprehensive coverage for a good evaluation of their strategy, the first steps of the SEZ have already generated work for both China and the local labour force. For example, the SEZ in Zambia, Nigeria and Ethiopia have, by 2013, generated approximately a combined 1849 Chinese and 11,192 local employment opportunities. It should be noted here that the 'Zone development in China [...] took at least a decade or longer to reach maturity' [46].

In the same vein, it is plausible that the SEZ in Tanzania will create employment and initiate aggregate demand growth. However, another question that may arise here is about what percentage of the unemployed or informally employed labour force would be employed in this new SEZ and whether that would be sufficient to ensure productivity and wage growth. Sustainable growth, driven by aggregate demand, furthermore, would depend on the type of jobs being offered. For stable and equitable growth, not only low-skilled jobs, but also middle-skilled and high-skilled jobs must be offered to the local population. With local employment in agriculture and fishing, there is a lack of skills among the local population to take up demanding industrial and management jobs. This void could possibly, to some extent, be filled by workers from other parts of Tanzania, mostly the nearby metropolis of Dar-es-Salaam. Education programmes for some of the 300,000 inhabitants of the Bagamoyo area would still be required to prevent rising social inequality due to income inequality and job polarization. Another reason why that is important is the danger – from a Tanzanian perspective – that too many of the created jobs will be taken by Chinese workers, as was common in previous projects.

Whether the development will be socially inclusive throughout its course is more questionable. Literature and news articles show evidence that within OBOR projects, being carried out outside of China, human rights are being neglected, locals face racism and that Chinese instead of local labour is favoured, resulting in more gains for the Chinese instead of a win–win situation [35]. This contradicts the initial goals as mentioned in the OBOR vision documents. However, it must be noted that these documents do not provide guidelines on how mutual learning, win–win situations, responsible development and so on should be accomplished. That would, however, be necessary if the goals are to be carried through all levels of involved actors.

Looking at the environment, there are reasons to be concerned about the preservation of the natural resources once the construction of the port starts and throughout its operation. Pressure by additional workers and more tourists, and most importantly, ocean vessels is likely to occur. Although no studies have shown the environmental impact in the vicinity to the PoB, there are studies from other parts of the world that have highlighted the exacerbating effects on the environment due to port operations [47]. Besides, with a community so reliant on occupations – characterized by its tradition – adhered to its natural environment, many questions can be raised regarding the impact of such a port on the natural environment and the local community: would it lead to sustainable prosperity? Although a new port may create many new job opportunities, it is unlikely that the local community will have the training and skills to adapt, and it is likely that the environment is affected negatively. Furthermore, global shipping of goods is a big contributor to emissions of carbon dioxide and other pollutants; local environmental deterioration due to port construction (area usage), leakage of oil and ballast water are almost inevitable [47]. Still, many activities in practice and research focus on 'sustainable ports' (Google finds 39,500 results). Such concepts, which have been applied elsewhere, for example, quite prominently in the Port of Rotterdam, would clearly contribute to the environmental sustainability of the port.

7.7 Conclusion and recommendations for further action

Despite all criticism, there is no doubt that the project entails huge potential for development; whether it will be sustainable on all three dimensions depends to a large extent on how it will be shaped. Social and environmental concerns need to be specifically addressed. Those issues are likely to affect the local residents of Bagamoyo, while the economic benefits of the port – most importantly, integration into the global economy and stronger ties within the region – affect a much larger area.

Local residents' concerns should play a larger role in the planning process. To ensure that this happens and that the good intentions specified in the OBOR goals are met, the project needs to be planned responsibly. A framework that could be used to shape the process is Responsible Research and Innovation (RRI) [48, 49], which can be seen as an extension of the concept of Sustainable Development that is more suitable as a framework for the design and development of technology [12]. The approach always includes a thorough consideration of involved stakeholders, which could greatly contribute to making the project beneficial for the local community.

The construction of a port in an ecologically diverse coastal zone will inevitably lead to environmental destruction through building the port on natural land. This is a trade-off that Tanzania is – quite understandably – willing to make, as a port with great regional potential is a very promising way to help its population out of poverty. To ensure its development towards a prosperous nation, environmental protection should not be forgotten. A compromise could be the establishment of ecological compensation areas in similar landscape forms, for example, further up along the coast. The set-up of such zones could even be made mandatory for investors who convert valuable nature into industrial zones.

With a rich environment comprising abundant natural resources and a potential labour force, yet unable to tap into these resources without the assistance of foreign powers is a weakness of East African countries. However, if African communities can influence their own changes by seizing the opportunities so offered by the Chinese through policy reforms and technological, societal and economic transcendence, the construction of the new port may bring prosperity that the people can 'lay down their melancholy' and reap the fruits of the new opportunities and influx of knowledge and technology into their country and area. Nevertheless, it is also worthwhile to find out whether Chinese investment will lead to sustainable prosperity or an undesirable situation where the people of Bagamoyo would have to 'lay down their hearts' because their land would not be the same due to exploitation by foreign powers.

7.8 Afterword: Limitations and further research

While the chapter paints a promising picture for the development of the Port of Bagamoyo and the Special Economic Zone project, it also shows there are reasons for concern. It is imperative, however, to critically reflect on the research work and opinions.

A limitation of paramount importance in this chapter is the scarcity of data, which severely limits the possibility of assessing the impacts the PoB project will have on the local populace. Additionally, any data on how cultural values of the local population will be taken into account are lacking, thus forcing speculation based on examples of countries in Tanzania's proximity a necessity. This might, therefore, not entirely hold and it is unclear what the extent is of China's and Tanzania's concern for the local citizens and their plans for incorporating their cultural values into the project. Another implication of the lack of data means that it cannot be verified what the plans for the natural environment are – as none could be found. This can either imply a lack of said plans or that these are present but not published yet. It remains to be seen what will happen in that regard.

Furthermore, any political factors have been ignored within this chapter as they were beyond the scope of the research. These factors play an important – if not the most important – role at an international level and could explain why any real cooperation between China and Western countries within the African continent has been absent so far. Political factors likely form major considerations from many countries along the Silk Route. The major complications these factors give are thus unaccounted for.

Further research should consider the above two major limitations of this work. To acquire more data, a field study is highly recommended as it will provide in-detail data on the exact handling of the local populace and the environment of Tanzania. Valuable data could be obtained to depict the situation of the port more accurately, which will help define whether or not the port project of Bagamoyo will lead to Sustainable Development, or better yet, what further actions would be best to further enhance the chance that the project will lead to Sustainable Development.

Additionally, it allows for a better understanding of sentiments across the local population, as these are currently scarce and oftentimes contradictory. Although the perception of the locals cannot serve as an adequate measure for challenges the project might face, it allows the identification of possible conflicts.

Lastly, a field study would allow the actors to determine a clear order in which the project should be scheduled and carried out and which steps to take in planning the project. This could allow for a better understanding of the sustainability implications of the project and how these have been accounted for by the Chinese and Tanzanians.

It could also be of interest to include a detailed evaluation of the political implications and their complications for the port project and how these might hamper, or enhance, the chance of Sustainable Development in the PoB project.

Acknowledgements

This chapter is based on a paper that the authors successfully concluded on the Technology and Global Development course at Delft University of Technology. Wim Ravesteijn, the module manager of this course, has thoroughly edited the text.

References

[1] Belt and Road Forum for International Cooperation, Vision and actions on jointly building Belt and Road. http://2017.beltandroadforum.org/english/n100/2017/0410/c22-45.html. Accessed on: 2 July 2021.

[2] Ministry of Finance of Tanzania, Tanzania's Five-year development plan 2016/17–2020/21. http://www.mof.go.tz/mofdocs/msemaji/Five%202016_17_2020_21.pdf, 2016.

[3] Tairo, A., Tanzania surrenders Bagamoyo port project to Chinese firm, 2017. http://www.theeastafrican.co.ke/business/Tanzania-Bagamoyo-port-project-to-Chinese/2560-4122244-rxa9wtz/index.html. Accessed on: 30 December 2017.

[4] FE Online, OBOR Troubles: How China's ambitious Silk Road initiative has hit massive potholes: From Pakistan to Tanzania. http://www.financialexpress.com/world-news/obor-troubles-how-chinas-ambitious-silk-road-initiative-has-hit-massive-potholes-from-pakistan-to-tanzania/1011158/. Accessed on: 22 January 2018.

[5] Office of the Leading Group for the Belt and Road Initiative, *Building the Belt and Road: Concept, Practice and China's Contribution*, Foreign Languages Press Co. Ltd., Beijing, 2017.

[6] Johnson, C.K., President Xi Jinping's 'Belt and Road' Initiative; A Practical Assessment of the Chinese Communist Party's Roadmap for China's Global Resurgence. Center for Strategic and International Studies, Washington, DC, 2016.

[7] Financial Times, China's Xi hails Belt and Road as 'project of the century,' 2017, *Financial Times*. https://www.ft.com/content/88d584a2-385e-11e7-821a-6027b8a20f23. Accessed on: 22 January 2018.

[8] Jinchen, T., 'One Belt and One Road': Connecting China and the World, 2016, McKinsey & Company; Capital Projects and Infrastructure. https://www.globalinfrastructureinitiative.com/article/one-belt-and-one-road-connecting-china-and-world. Accessed on: 2 July 2021.

[9] Verlare, J. & Putten, P. van der, 'One Belt, One Road': An Opportunity for the EU's Security Strategy, Netherlands Institute of International Relations, Clingendael, 2015.

[10] WCED (World Commission on Environment and Development), Report of the World Commission on Environment and Development: Our Common Future, http://www.un-documents.net/our-common-future.pdf, 1987. Accessed on: 21 May 2021.

[11] United Nations, Transforming our world: The 2030 Agenda for Sustainable Development, 2016. https://sustainabledevelopment.un.org/post2015/transformingourworld. Accessed on: 22 January 2018.

[12] Ravesteijn, W., He, J. & Chen, C., Responsible innovation and stakeholder management in infrastructures: The Nansha Port Railway Project. *Ocean & Coastal Management*, 100, pp. 1–9, 2014.

[13] Dwarakish, G. S. & Salim, A. M., Review on the role of ports in the development of a nation. *Aquatic Procedia*, 4, pp. 295–301, 2015.

[14] Bagamoyo: Throw down your heart, 2012. https://hearabout.wordpress.com/2012/08/08/bagamoyo-throw-down-your-heart/. Accessed on: 30 December 2017.

[15] Fabian, S., East Africa's Gorée: Slave trade and slave tourism in Bagamoyo, Tanzania. *Canadian Journal of African Studies/Revue Canadienne des Études Africaines*, 47(1), 95–114, 2013.

[16] Campbell, G., The East African slave trade, 1861–1895: The 'Southern' complex. *The International Journal of African Historical Studies*, 22(1), pp. 1–26, 1989.

[17] Myers, J., Which are Africa's fastest growing economies? 2016. https://www.weforum.org/agenda/2016/04/which-are-africa-s-fastest-growing-economies/. Accessed on: 20 January 2018.

[18] Korybko, A., Hybrid Wars 8. Lots of trouble to Tanzania, *Oriental-Review.org*. https://orientalreview.org/2017/01/20/hybrid-wars-8-lots-of-trouble-to-tanzania/. Accessed on: 30 December 2017.

[19] PWC, (no title), https://www.pwc.com/gx/en/transportation-logistics/publications/africa-infrastructure-investment/assets/tanzania.pdf, n.d. Accessed on: 21 May 2021.

[20] Odhiambo, N.M., Financial deepening, capital inflows and economic growth nexus in Tanzania: A multivariate model. *Journal of Social Sciences*, 28(1), pp. 65–71, 2011.

[21] Sosovele, H., Bagamoyo area: Economic and social briefing, http://encapafrica.org/, n.d.

[22] Haralambides, H., Veldman, S., Drunen, E. van & Liu, M., Determinants of a regional port-centric logistics hub: The case of East Africa. *Maritime Economics & Logistics*, 13(1), pp. 78–97, 2011.

[23] Mabeya, C., New port set to launch major economic boost, 2017. http://www.chinadaily.com.cn/a/201711/10/WS5a2927baa310fcb6fafd417a.html. Accessed on: 2 July 2021.

[24] Lam, J.S.L. & Notteboom, T., The greening of ports: A comparison of port management tools used by leading ports in Asia and Europe. *Transport Reviews*, 34(2), pp. 169–189, 2014.

[25] Sustainability, 2015. https://www.portofrotterdam.com/en/the-port/sustainability. Accessed on: 2 January 2018.

[26] Dreher, A., Fuchs, A., Parks, B., Strange, A. & Tierney, M., Apples and dragon fruits: The determinants of aid and other forms of state financing from China to Africa, 2016. https://papers.ssrn.com/abstract=2855935. Accessed on: 22 January 2018.

[27] Easterly, W., *The White Man's Burden*. London: Penguin Books, 2006.

[28] Moyo, D., *Dead Aid*. London: Penguin Books, 2009.

[29] Bräutigam, D., *The Dragon's Gift*. Oxford: Oxford University Press, 2009.

[30] Ramo, J.C., *The Beijing Consensus*. London: Foreign Policy Centre, 2004.

[31] Yang, Y., Beijing Consensus or Washington Consensus: What explains China's success? Development Outreach, pp. 28–30, World Bank, 2011. https://openknowledge.worldbank.org/bitstream/handle/10986/6098/deor_13_1_26.pdf?sequence=2&isAllowed=y. Accessed on: 21 May 2021.

[32] World Foresight Forum, WFF02 Issue Brief: The Beijing Consensus, 2011.

[33] Dirlik, A., Beijing Consensus: Beijing 'Gongshi.' Who Recognizes Whom and to What End? Presented at the Globalization and Autonomy Online Compendium, 2006.

[34] Lekorwe, M., Chingwete, A., Okuru, M. & Samson, R., China's Growing Presence in Africa Wins Largely Positive Popular Reviews, Afrobarometer No. 122, 2016.

[35] Wang, F.-L. & Elliot, E.A., China in Africa: Presence, perceptions and prospects. *Journal of Contemporary China*, 23(90), pp. 1012–32, 2014.

[36] Geerts, S., Xinwa, N. & Rossouw, D., Africans' perceptions of Chinese business in Africa: A survey, 2014. http://www.globethics.net/documents/4289936/13403252/GE_Focus_18_web_1.pdf. Accessed on: 21 May 2021.

[37] Kelly, G., Brazys, S. & Elkink, J.A., *The Dragon's Curse? China, the World Bank, and Perceptions of Corruption in Tanzania*. AidData Working Paper #26. Williamsburg, VA: AidData at William & Mary, 2016. http://docs.aiddata.org/ad4/files/wps26_the_dragons_curse.pdf. Accessed on: 21 May 2021.

[38] ISSAfrica.org., Ghana takes action against illegal Chinese miners, 2013, https://issafrica.org/iss-today/ghana-takes-action-against-illegal-chinese-miners. Accessed on: 28 December 2017.

[39] Veeck, G. & Diop, S.H.A., Chinese engagement with Africa: The case of Madagascar. *Eurasian Geography and Economics*, 53, pp. 400–418, 2012.

[40] Dixon, R., Chinese officials accused of smuggling African ivory on official visits. *Los Angeles Times*, 2014. http://www.latimes.com/world/africa/la-fg-africa-china-ivory-smuggling-20141106-story.html. Accessed on: 21 May 2021.

[41] Rotterdam Climate Initiative, RCI-ZEROCO2, 2011. http://www.zeroco2.no/projects/rotterdam-climate-initiative. Accessed on: 21 May 2021.

[42] Onnis, B., China in Africa: Challenges and opportunities. In: Beretta, S., Berkofsky, A. & Zhang, L. (eds), *Understanding China Today*, pp. 129–148. Cham: Springer, 2017.

[43] Yap, W.Y. & Lam, J.S.L., 80 million-twenty-foot-equivalent-unit container port? Sustainability issues in port and coastal development. *Ocean & Coastal Management*, 71 (Supplement C), pp. 13–25, 2013.

[44] Yam, S., Influx of Chinese transforms the landscape of Madagascar, 2018. http://www.scmp.com/business/article/1300808/influx-chinese-transforms-landscape-madagascar. Accessed on: 21 May 2021.

[45] Pew Research Centre, Chapter 2: China's Image, 2014. http://www.pewglobal.org/2014/07/14/chapter-2-chinas-image/. Accessed on: 21 May 2021.

[46] Bräutigam, D. & Tang, X., 'Going global in groups': Structural transformation and China's special economic zones overseas. *World Development*, 63, pp. 78–91, 2014.

[47] Ng, A.K.Y. & Song, S., The environmental impacts of pollutants generated by routine shipping operations on ports. *Ocean & Coastal Management*, 53(5–6), pp. 301–311, 2010.

[48] Owen, R., Macnaghten, P. & Stilgoe, J., Responsible research and innovation: From science in society to science for society, with society. *Science and Public Policy*, 39(6), pp. 751–760, 2012.

[49] Stilgoe, J., Owen, R. & Macnaghten, P., Developing a framework for responsible innovation. *Research Policy*, 42(9), pp. 1568–1580, 2013.

Part III

Responsible Port Innovation: Means and methods

CHAPTER 8

The Dalian Port Cluster: Spatial practice of the One Belt, One Road initiative and potential opportunities of the methodology of Responsible Research and Innovation

Penglin Zhu & Carola Hein

Department of History of Architecture and Urban Planning, Delft University of Technology, The Netherlands

Abstract

The Chinese government is heavily investing in trade-related infrastructure with its One Belt, One Road (OBOR) initiative – a proposal to build a New Silk Road or network of trade routes. This infrastructure-driven innovation will have a large economic, spatial and environmental impact on China; it is also likely to transform entire regions throughout Asia, Africa and Europe. In pursuing it, the Chinese government seeks to meaningfully combine a strong national interest in infrastructure and economic development with an equally strong national interest in environmental sustainability. This chapter explores how it might meet both goals. How do the initiative's many projects intersect with local interests and needs in China? Will these investments carry out the proclaimed desire of the Chinese government for environmentally Sustainable Development? This chapter examines these questions by looking at how various governmental players are translating these investments and proposals into spatial plans. It focuses on port city regions, using the Dalian region in northeast China, one of the OBOR hubs, as an example. Port regions are key spatial components of the OBOR initiative: as infrastructural hubs and centres for trade and innovation, as sites of national oil storage or hubs for the development of new maritime technology. As places where the industrial production and long-distance infrastructure of the OBOR initiative encounter densely populated areas, they are also particularly at risk from climate change, including rising sea levels. Thus, the OBOR initiative's goals of economic improvement and environmental sustainability may well collide. In Dalian, the chapter tests the methodology of Responsible Research and Innovation (RRI) – a tool developed in Europe to assess research and products in terms of social and environmental impact and transparency – to see how it could provide guidance for future development within the OBOR initiative that achieves multi-faceted goals.

Keywords: One Belt and One Road, planning, economy, environment, Responsible Research and Innovation, Dalian, port cluster

8.1 Introduction

Since announcing the OBOR initiative [1, 2] in late 2013, the Chinese government has promised to invest some US$1 trillion in massive land-based infrastructure (roads, railways, power plants and pipelines) along the historic Silk Route, and in port and maritime planning on the south side of the country, the so-called Maritime Silk Road. It will soon invest in the economic belt of the Northern Ice Sea Route as it opens up due to global warming [3]. This enormous investment is likely to transform cities and landscapes in Asia, the Middle East and parts of the European Union. Ports, port cities, and regions of neighbouring areas – specifically the biggest ports with the most extensive infrastructure in China, including Shanghai, Guangzhou, Tianjin, Dalian and Haikou – are key economic and spatial nodes in the OBOR initiative. These port cities are land- and sea-based hubs for commodities headed to international destinations. They are also hubs for the production and transportation of energy and new maritime technologies, often hosting extensive industries (including many petroleum-based ones).

Current Chinese national economic plans and spatial strategies have two objectives: carrying out national demands for economic development and ecological sustainability and constraining what they frame as unhealthy autonomous local development. Economic development and ecological sustainability can be opposing objectives. Port industries are often already big polluters. Located on lowlands, port cities and their coastal regions are particularly at risk of water-related environmental catastrophes [4, 5]. The OBOR initiative promises to further accelerate urbanization and industrialization, squeezing commerce, people and infrastructure into the already fragile ecosystem of the Chinese coastline. The dangers of this development have become evident in recent environmental disasters in port cities, including the 2015 explosions in the Tianjin Port region and a recent oil spill in the East China Sea in January 2018 [6]. Direct economic losses from marine disasters have totalled more than US$32.8 billion [7].

This chapter explores how the OBOR initiative aims to balance development and sustainability through spatial plans at diverse territorial scales – from provincial to municipal or port areas. Subsequently, it examines how the national government has coordinated provincial governments, local municipalities and various industrial stakeholders to create regional synergetic development and spatial strategies to realize the OBOR concept. It studies how and to what degree the various institutions have implemented spatial strategies. It specifically focuses on communication between the government, industrial representatives and the general public to assess the degree of transparency in planning.

The objective of balancing economic with ecological sustainability is new to the OBOR initiative, but the respective strategies are based on previous economic policies or concepts of environmental sustainability. Old and new plans and strategies have already shaped the spatial structure of ports, port cities and their neighbouring areas across the country. The OBOR initiative, the 13th Five-Year Plan of 2016 and the programme of Free Trade Zones (FTZs), which since 2013 have eased economic regulations on industrial and commercial activities, including foreign ones, have inherited concepts of economic development and ecological sustainability from previous national strategies [8], such as the Plan for Revitalizing the 'Old Industrial Bases in Northeast China' [9] issued by the State Council in 2006. They also inherited the principle of environmental protection from the State Council's Decision on Implementing the Scientific Outlook on Development and Strengthening Environmental Protection issued in 2005 [10].

The chapter contends that the OBOR initiative imagines and effectively establishes a new spatial scale – what we call the regional *port cluster* – at which it will implement its policy. The port cluster, conceived here both physically and analytically, includes a port, its city, the FTZ and their neighbouring area. It is first a regional space that supports port activities and comprises port facilities, maritime-related administrative offices and research institutions, logistics centres and infrastructural links of roads, rail and airports. In the absence of a single administrative unit in charge of this physical space, it is governed by various provincial, municipal and port institutions with often different interests and powers; in effect, it is a conceptual space. The port and city of Dalian, the Liaoning FTZ and the neighbouring areas are the case study of the chapter due to their representative advantages and risks in the regional spatial plans for the OBOR initiative. Dalian Port, Dalian City and Liaoning Province are important spatial nodes in both the Economic Belt and Economic Road. Chosen by the central government to host the OBOR initiative's activities, they have received national policy support and large investments from the central government. The chapter conceives the port and city of Dalian, the Liaoning FTZ and three provinces of Liaoning, Jilin and Heilongjiang as the *Dalian Port Cluster*.

Finally, the chapter explores the research hypotheses in the case study of the Dalian Port Cluster using the methodology of Responsible Research and Innovation (RRI). It imagines how applying the mechanisms of RRI could improve the planning process and regional implementations of the OBOR initiative. Moreover, how RRI can improve transparency in planning, which is considered a key element in guaranteeing responsible planning. Developed by scholars, including the Dutch philosopher René Von Schomberg from the Directorate General for Research at the European Commission, this methodology has multiple goals. RRI names five mechanisms to help governments and urban planners assess research and innovation: 'technology assessment and foresight'; 'application of the precautionary'; 'normative/ethical principles to design technology'; 'innovation governance'; and 'stakeholder involvement and public engagement'. RRI conceives the normative principles of 'ethically acceptable', 'sustainable', 'socially desirable' and 'inclusive', involving the collaboration of innovators and social actors. That is, RRI helps scholars and professionals discern whether research or new products benefit society and the environment, asking if the creators of that research or products have already involved diverse societal groups, if researchers and companies have used transparent and open processes to make the planning more reliable and if they have assessed risks and opportunities (social, ethical, environmental, technical and commercial) [11].

Though RRI has mostly only been used in the European Union, where social, economic and political systems are very different from those in China, it has gradually attracted global scholarly attention. In fact, Chinese economists Yang and Han have already used RRI to analyse new Chinese economic policies [12]. They have argued that the mechanisms of Chinese stakeholder participation need to be enhanced to optimize economic innovation. And Chinese policymakers and think tanks are keen to know more about the approach of RRI. Its assessments, especially of environmental policies, can improve Chinese policymaking and execution both within China and in OBOR initiative-related construction abroad.

8.2 Literature review

Scholars have started to point out the effect of this construction on the built environment – including Troy Sternberg, a specialist on the interaction of natural hazards with societies and the environment in the Gobi area in China – in order to create awareness of the unprecedented future transformation of the region's built environment in Central Asia [13]. Numerous scholars have studied the environmental advantages and risks of the OBOR initiative's spatial projects in various border areas, including China–Russia [14], China–EU [15] and China–Iran [16]. Some have started to notice changes in port infrastructure and facilities; Orestis Schinas, professor of shipping and shipping finance at the Hamburg School of Business Administration, argues that the OBOR initiative has influenced existing services and caused the re-routing of commodity flows [17]. Ports and port cities are priorities of the OBOR initiative, and Prem Chhetri, professor of geo-logistics from the University of Melbourne, and Paul Tae-Woo Lee, professor of marine economics from Zhejiang University, argue for studying these sites as cluster-led global port logistics to understand how these logistic networks have been strategically aligned in Asia by the

OBOR initiative, and they also need to be understood functionally as global trading hubs and transport corridors [18], moreover, as focal points for socio-economic-political interests that drive the economic growth and spatial development in the context of the OBOR initiative [19].

Official planning coordinates the enormous spatial projects of the OBOR initiative. Scholars have recognized its role in promoting economic, technological, political and social innovation in light of the initiative. But there is a lack of conversation about planning the world's largest port infrastructure and its effect on the environment. This chapter addresses this significant lacuna in this conversation. It highlights the importance of regional-scale planning of the port, the port city and neighbouring areas for achieving the economic and environmental goals of the OBOR initiative. It looks beyond the individual port, or a group of neighbouring area ports, to consider the relationship of the port to its vast hinterlands.

The OBOR project focuses on developing a number of regions with strong economies, wide-ranging trade relations, long histories and distinctive governance structures. Such developments require an appropriate conceptual and analytical framework. We argue that the Chinese practices are working at the level of what we propose is a *port cluster*. While this term is not mentioned in the Chinese texts, its conceptualization is necessary to understand the various actors and spaces involved and needed for and affected by the function of the port and the transformation of the sea at a regional level. Considering a larger region as the site in which to assess, guide and plan port city processes can help politicians, institutions and planners rethink competitiveness and synergy among cities as they consider sharing innovative information. It is especially important to understand what actors are involved at what spatial scale and what tools they have available.

Our definition of the port cluster enriches the common definition of the port cluster, which describes a group of neighbouring ports in competitions. For example, Peter. W. de Langen, a specialist in port logistics, argues that ports can be analysed as a group constituted by economic activities related to shipping routines; however, he has not studied the importance of the neighbouring area to the development of the port and port city [20]. In our usage, a port cluster includes the port, port cities and hinterland based on geographical distributions of an inland port in China, such as the division of three Chinese port clusters (the northeast cluster, exclusively invested by the port of Dalian, the central cluster led by the port of Tianjin and the southeast cluster cooperating with the ports of Ningbo and Xiamen) by maritime logistics expert Jason Monios from Edinburgh Napier University [21].

The chapter argues that the port cluster is the appropriate spatial concept for studying the spatial development of the port and city of Dalian and the neighbouring areas as sites of the OBOR initiative. It can illuminate how the OBOR initiative is re-branding the ports and their hinterlands as regions labelled with balancing economic development and ecological sustainability. Scholars have used this concept of branding cities to study the arrangement of global cities [22], urban regeneration [23], spatial strategies of significant events [24] and urban governance in China [25]. The idea of the regional port cluster enriches the research on city branding in China, which focuses on service- and innovation-oriented cities, such as the Pearl River Delta area [26] and the Jing-Jin-Ji area [27], while rarely discussing the outdated resource-based and heavy industry-based cities, including the ports and hinterlands in the three northeastern provinces of Liaoning, Jilin and Heilongjiang. Furthermore, it is important to study the environmental planning of the OBOR initiative. For example, Martin de Jong, a Dutch specialist in urban and infrastructure development in China, uses the city branding concept to study the ecological modernization of various city groups in northeast China [28, 29].

8.3 Planning in the Dalian port region: Towards a balance of economic development and ecological sustainability?

The Chinese government has historically carried a concern for ecological sustainability. Together with the Ministry of Foreign Affairs, the National Development and Reform Commission and the Ministry of Commerce, the Ministry of Ecology and Environment issued Guidance on Promoting the Construction of the Green 'Belt and Road' (关于推进绿色"一带一路"建设的指导意见) in April 2017 [30] and the Belt and Road Ecological and Environmental Cooperation Plan in May 2017 [31], emphasizing the national concern for ecological sustainability in the OBOR's projects. The Outline of the 13th Five-Year Plan for the National Economic and Social Development of the People's Republic of China (13th Five-Year Plan) (中华人民共和国国民经济和社会发展第十三个五年规划纲要) [32], issued in 2016, highlights ecological sustainability in one chapter, 'Moving faster to optimize the ecology and environment', as guidance for local economic plans.

But the national government's practice to balance economic development with regional ecological sustainability so far has had limited success in the context of the OBOR initiative. For example, the Chinese Territory Development Plan (主体功能区规划) [33] and the Chinese Marine Development Plan [34] issued by the State Council in 2011 and 2015 respectively, have categorized four types of territories to develop: 'development zones to be optimized', 'key development zones', 'limited development zone' and 'prohibited development zones'. These plans do not explore how to balance preservation with the coastline petroleum industry, an economic pillar of Chinese port cities such as Qingdao and Dalian. Meanwhile, the developing environment production industry, which is part of ecological sustainability strategies, has not repaid the economic increase. According to Xiuyan Sun, a journalist at the *People's Daily*, the environmental protection industry is facing difficulties since the annual production value is less than 2% of the national GDP, and it is not a backbone industry [35].

To fully realize the economic and ecological goals of the OBOR initiative, the Chinese government needs regional-scale planning agents. The Chinese government has long understood the importance of regional planning. Mostly, these regional plans identified economic and technical strategies. Specifically, regional government entities have only briefly existed, such as between 1949 and 1953, shown to the left of Figure 8.1. Nonetheless, national economic plans have addressed regional scales, old industrial bases in northern China (since 2006), the Bohai Economic Rim (since 1992) and the Pearl Economic Delta (since 1994). The Vision for Maritime Cooperation under the Belt and Road Initiative [36], a 2017 document by the National Development and Reform Commission and the State Oceanic Administration ['一带一路'建设海上合作设想] and the 13th Five-Year Plan each mapped out an overall strategy for the coordinated development of city groups of northeast China and central China. Effectively, these plans established a vision of a comprehensive regional plan that would protect regional marine environments and address climate change. To achieve these, it would promote international cooperation, particularly to develop so-called blue carbon, a mechanism for using marine activities and marine organisms to absorb carbon dioxide from the atmosphere, fix it and store it in the ocean.

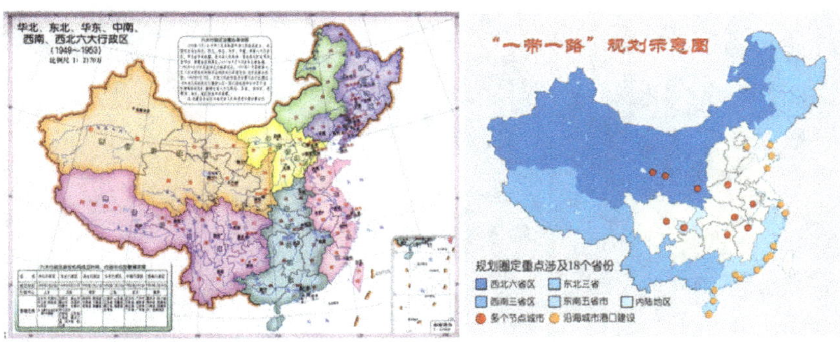

Figure 8.1: On the left, the map of the regional governments (六大行政区); on the right, regional set-up conceived in the OBOR initiative based on five regions shown in blue. Several regions such as northeast China are inherited from the historical concept. (Source: on the left: Sinomaps 1953; on the right: Souhu Business, http://business.sohu.com/20150405/n410831540.shtml.)

The national government plans on a regional scale in what we call the Dalian Port Cluster in northeast China includes the provinces of Liaoning, Jilin and Heilongjiang (Figure 8.2). The port and city of Dalian and the Liaoning FTZ are major parts of the Dalian Port Cluster, while the Jinlin and Heilongjiang provinces are regional hinterlands in the cluster. Northeast China serves as an extension of the new Maritime Silk Route that connects the Middle East, East Asia and Africa. Liaoning province serves as the starting point of the OBOR infrastructure, and Jilin and Heilong provinces are important nodes on the Belt, whose massive markets are attractive to international businesses. The Territory Development Plan and the Marine Development Plan have imagined Dalian, Shenyang, capital of the province, and other parts of Liaoning province as a Chinese international shipping centre in northeast Asia and an international logistics centre of the OBOR initiative. The mentioned two plans designate part of the Liaoning province as the Northeast International Logistics Corridor connecting the major industrial cities in northeast China. The 13th Five-Year Plan for Revitalizing the Industrial Bases in Northeast China issued by the National Development and Reforming Commission in 2017, presents

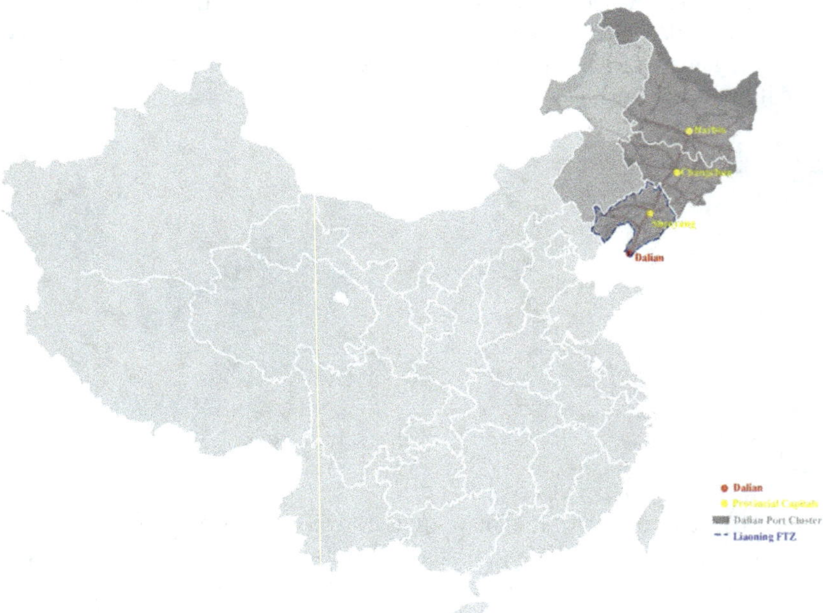

Figure 8.2: Schematic of the Dalian Port Cluster. (Source: Penglin Zhu.)

the economic connection between Liaoning province (orange) and Jilin and Heilongjiang provinces (green), showing the importance of Liaoning province, specifically, the strategical position of the port of Dalian. The plan proposes an infrastructural and developmental axis connecting the northeastern cities, including Harbin (capital of Heilongjiang province), Changchun (capital of Jilin province), Shenyang and Dalian [37] – the purple double-headed arrow in Figure 8.3. Within this axis, infrastructure links Dalian Port to the river port in Harbin, a strategic location for energy transportation and storage in the northeast, and further on to the Russian ports of Khabarovsk, Blagoveshchensk and Nikolayevsk [38]. These proposed infrastructural networks are also part of Medium- and Long-Term Plans for the National Gas Pipeline Network (中长期油气管网规划), issued by the National Development and Reform Commission and National Energy Administration in 2017, which proposes to construct storage for liquified natural gas (LNG), reconstruct and extend the oil port and develop a new complementary pipeline network connecting offshore oil and gas platform to the hinterland [39].

Despite the national government's focus on planning on a regional scale here, there are no regional governmental institutions administrating the Dalian Port Cluster. Yet to actually build anything, the OBOR initiative requires the support and collaboration of provincial governments, local municipalities, companies, nationally funded think tanks and academies, petroleum companies active in the ports and other industrial companies. In particular, port development requires investment into the landside infrastructure for distribution on a regional scale [40]. The development of a smooth and efficient transportation route here will rely on strong support from various provincial and municipal governments in the region that are in charge of improving port infrastructure, strengthening port service and developing maritime logistics. The national government, provincial governments and local municipalities need to create synergies among several initiatives, the development of the port, the creation of a special economic area that facilitates (foreign) trade and regional-scale infrastructural planning. Local developing plans for 2016–2020 in each of three provinces affected by the OBOR project – Heilongjiang province, Jilin province and Liaoning province – do mention the OBOR initiative as a priority. For instance, the 13th Five-Year Plan for the Economic and Social Development of Liaoning Province, issued by the People's Commune of Liaoning Province in 2016, proposes to use the port of Dalian as a keystone to further build Liaoning province as the spatial node connecting the EU and East Asia [41].

But the absence of regional governmental institutions makes it difficult for any regional synergy to develop. Provincial and local governments tend to ignore parts of the national plans to pursue their own interests. Administrators of these institutions are not always in agreement with the OBOR initiative's

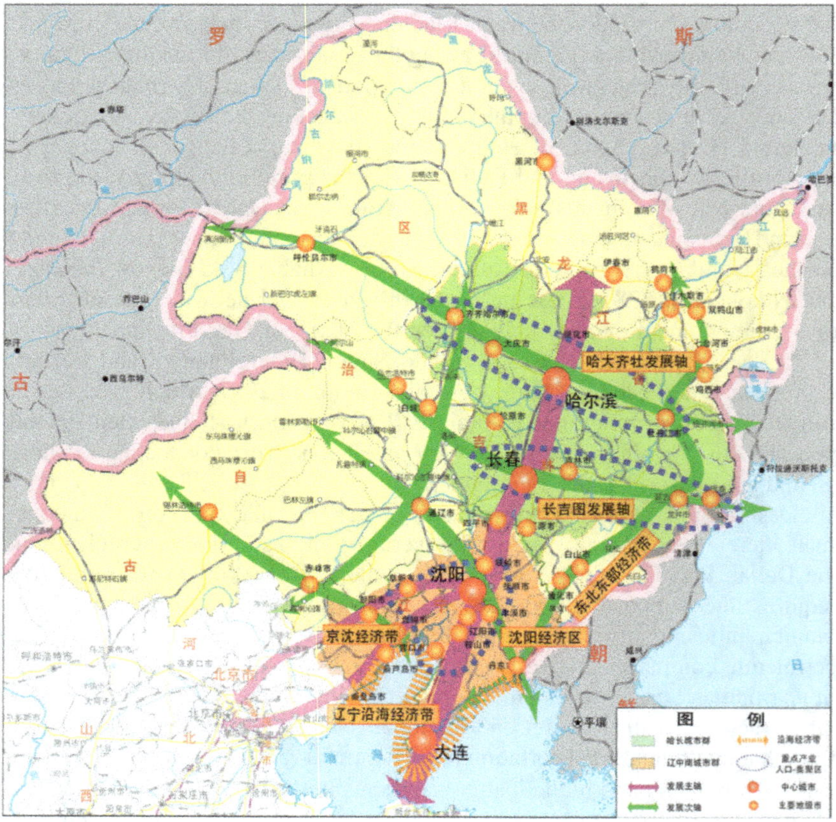

Figure 8.3: Thirteenth Five-Year Plan for Revitalizing the Old Schematic Plan of the cities and region identified in the 13th Five-Year Plan for Revitalizing the Old Industrial Bases in northeast China, 十三五东北地区振兴计划. (Source: National Development and Reform Commission, http://www.ndrc.gov.cn/zcfb/zcfbghwb/201612/W020161219594214175674.pdf).

overarching goals or specific plans. Some try to create plans that realize national strategies while also furthering local goals, while others are primarily keen to capture part of the enormous national investment for local or other interests. In the Dalian Port Cluster, provincial plans thus far have focused on developing new green environmental protection technologies to meet both economic and ecological objectives. A positive example is the Plan of the Marine Development of Liaoning Province in 2017, which designates the Zhuangheishi area as a resource-saving and technology-oriented industrial area, including industrial-scale wind and tide energy generation and resource-saving [42]. However, the 2017 Revised Comprehensive Urban Plan of Dalian (2001–2020) focuses on economic development and only mentions ecological sustainability in the programme description of the transportation infrastructure [43].

Even when the various administrative levels mention overarching goals, Liaoning has not addressed how to coordinate its spatial plan for OBOR with those of Jilin province and Heilongjiang province, and neither of those provinces have proposed any communication with the others in their provincial 13th Five-Year Plans. Inter-province (or inter-municipal) communication channels and transparent procedures are missing, and processes to present the planning progress to the general public are absent.

Support for the implementation of the OBOR plan also needs to come from other stakeholders, such as the state-owned Dalian Port Company, but again it is unclear how they will balance economic and environmental goals and how they will translate these into spatial plans. Although the Dalian Port Company issued a Plan towards the Green Ecological Port for 2015–2020 in 2014, which includes numerous environmentally friendly concepts – such as low carbon, environmentally friendly loading and unloading transportation equipment, solar energy for cleaning facilities, green building technologies, increasing green land in the port and new oil spill emergency and prevention systems [44] – it is not yet clear whether and how these will play out in port construction or if the company will make other changes in response to the OBOR plan.

The mystery of planning, progress and local supervision is the primary barrier to realizing the OBOR initiative in the Dalian Port Cluster. Without transparency, the various stakeholders and the general public do not know to what degree the governments have executed their promises. Liaoning province has planned to build China's largest crude oil terminals (at 450,000 tonnes and 300,000 tonnes, plus a 400,000-tonne ore terminal and several 200,000-class container terminals) in Dalian's Dayou Bay, but it has not clarified how it will both boost its annual GDP and protect the fragile coastal environment. The Dalian Port Company Limited has claimed to use the newest protection system for oil storage tanks since 2016 [45], but several fires have broken out there, the most recent in August 2017 [46]. The port company has not explained to the general public to what degree the company can complete its promise. Information on other shortcomings is similarly missing. For instance, according to a report from the Third Environmental Protection Inspectorate from the central government [47], the Dalian Port Company had illegally reclaimed 585 hectares of land from the water during the construction of the Taiping Bay Port area. The inspectorate from the national government is touring domestically and can only defend the national interests rhetorically since it is not a law enforcement authority. The general public still does not know why the local supervision agency, which does have that authority, has ignored these actions of the Dalian Port Company.

Truly effective communication guarantees all parties that goals are achieved, that national and local interests are aligned, and that planning is handled efficiently. It includes options for public engagement and public debate on policy, with the real possibility that officials will change the policy in response to public comments. Chinese institutions have addressed the need for establishing a communication platform. The national 2007 Urban and Rural Planning Code of the People's Republic of China designed a way for people to communicate but will not be a permanent or convenient tool for the general public for a long time [48]. The current Chinese planning system offers some channels through which the general public can send in comments or complaints on urban plans, including petitioning, but people have only briefly used this tool, and it does not obligate officials to respond. For example, the Dalian Port Company posted the first and second versions of the Environmental Impact Assessment of the Dalian Port Taipingwan Area Master Plan on its website in 2017 for ten days, allowing readers to comment only via telephone or email or by going in person to the Dalian Port Company offices, not online [49, 50]. Whether the planners responded to people or modified their plans, thus far, has not been announced on social media. A practice of negotiation or debate is currently not fully established in China. Transparent and straight communication among the governments, various stakeholders and the general public is needed to realize the overall goal of balancing economic and environmental objectives.

In short, the official stakeholders who execute national strategies and translate them into spatial plans have only partially adopted the national government's environmental and economic goals. How can the Chinese national government, provincial governments and the local municipality improve spatial planning and implementation and achieve their national goals in the Dalian Port Cluster? What are the barriers that they need to overcome?

8.4 Responsible Research and Innovation: A potential tool for optimizing the planning processes of the Dalian Port Cluster

The chapter argues that the RRI methodology developed in Europe can help Chinese regional governments develop transparent processes and effective communication in order to realize the national OBOR plans. In particular, it can help policymakers address systems in China and Europe in which spatial planning is a top-down practice with limited, albeit carefully legislated, public participation. The European Commission has used RRI to evaluate programmes in economics and applied sciences, and to address ethical questions specific to technology [51, 52]. More importantly for our purposes, it has used RRI to evaluate potential impacts of proposed interventions in environment and society, including multi-actor and public engagement in research and innovation in the European Framework Programmes, such as the Horizon 2020 EU Research and Innovation programme [53]. The methodology of RRI promotes openness and transparency to help both scholars and administrators assess environmental sustainability and societal desirability of any innovation that has input from multiple stakeholders [54]. It fosters transparent communication, in which 'societal actors and innovators become mutual[ly] responsive to each other' [55]; such communication can help the various innovators in the built environment – governments, urban planners and industry representatives, and the general public – to optimize current planning, making it more reliable and acceptable to society and reducing its risks. RRI can help stakeholders better balance economic development and ecological sustainability in regional-scale planning in the Dalian Port Cluster.

This section uses the aforementioned five mechanisms to assess China's current national policies and spatial plans, and their implementation in the Dalian Port Cluster. The assessments can outline how the government can better balance its two objectives in the Dalian Port Cluster. In turn, using the five mechanisms to assess OBOR projects enriches the methodology of RRI, helping us to refine the methodology.

8.4.1 Technology assessment and foresight

This mechanism can help governments and industrial stakeholders anticipate the positive and negative impacts of OBOR's two objectives – environmental and economic. It posits that the settings of research priorities and the anticipated impacts need to be subjected to societal review. It means that the proposals from scientists (scholars from the national academies and national think tanks) should be extended to include an assessment of societal impacts. It aims to reduce the human cost of trial and error, and benefits from the social learning process of stakeholders and technological innovators.

So far, governmental institutions have not assessed societal impact. The official website of the National Natural Science Foundation of China (NSFC), the national academic institution affiliated with the State Council that forms national strategies and plans to develop science and technology [56], only mentions one international seminar for the port development in Tianjin in 2017 [57]. Specifically, the NSFC has not fully implemented the promises issued in the Cooperation Action Plan of the National Science Foundation of China to Build One Belt One Road Initiative in 2017 [58], in which the

NSFC claimed to organize seminars and conferences and establish research centres for scholars from the countries participating in the OBOR to balance research on regional economic development, environmental preservation and green technology. The NSFC needs to not only complete its promises to support the OBOR's projects but should extend the current programme and organize more global conferences and seminars, not only for scholars but for the general public as well.

8.4.2 The precautionary principle

The precautionary principle mechanism is a tool that can help the national government, provincial governments and local municipalities make sustainable spatial plans to avoid economic and ecological disasters in the Dalian Port Cluster. It posits that stakeholders have the desire to prevent harm or to diminish the harm of potential innovation. The mechanism aims to cope with possible risks where scientific understanding is yet incomplete. Considering historical context can help people assess the safety and the sustainability of new technologies. In the case of the Dalian Port Cluster, applying the precautionary principle could stop innovations that might have negative spatial or technological impacts. For instance, if any construction of the spatial projects of the OBOR initiative in the Dalian Port Cluster might harm the health or interests of the locals, the national government and the three provincial governments could refer to historical experiences of similar cases. Neither a policy of historical review nor a mechanism for speedy reaction currently seems to be part of the OBOR initiative. Thus, a careful review of historical precedents of regional-scale economic transformation is urgently required to prevent any harm. Moreover, the national government should establish a fast-reacting mechanism with which the three provinces can coordinate easily to terminate negative projects immediately.

8.4.3 Normative/ethical principles for designing technology

The mechanism of normative/ethical principles for designing technology engages the barrier of un-transparent processes. In particular, it aims to incorporate the ethical principles of safety and privacy into the planning processes that can lead to well-accepted planning advances in the Dalian Port Cluster. Governments cannot guarantee the safety and privacy of societal actors without a transparent planning process. The OBOR initiative has addressed the notion of safety superficially, only mentioning in spatial plans the marine challenges of climate change and the security of marine transportation [59], while not addressing the environmental risks of massive oil storage in Dalian Port. Moreover, the plans and policies of the OBOR initiative have not safeguarded the privacy of the general public. Transparency is a long-standing and critical topic in China. The Chinese government needs a long time to plan, execute and supervise the OBOR, during which it must consider environmental risks and address transparency and privacy.

8.4.4 Innovation governance and stakeholder involvement

The mechanism of *innovation governance* and *stakeholder involvement* assesses collaboration among the various stakeholders, asking how efficient it is. It posits that good communication helps make the creation and execution of plans more effective. It includes three aspects of governance innovations: multi-stakeholder involvement; use of codes of conduct; and adoption of standards, certification and self-regulation. The aspect of multi-stakeholder involvement aims to bring together actors from industry, society and academy to define plans for the responsible development of the Dalian Port Cluster. Bringing various actors together can help them take up a position on all perspectives instead of individually focusing on specific points. The aspect of the use of codes of conduct allows a proactive and independent scientific community to constructively steer the planning process of OBOR projects, identifying risks and benefits at an early stage and reporting on them to public authorities. The aspect of the adoption of standards, certification and self-regulation are fundamental requirements to allow for responsible development.

Currently, those stakeholders mainly follow the government or legally have to obey its guidance. Their subjective impact on the OBOR's projects has been negligible. Although the State Council has approved the concept of 'fate of the human community' [60], which implies the importance of the OBOR initiative, in the new constitution, there are no examples demonstrating that the constitution and legacy can be refined according to the standards and certifications of the stakeholders. Thus, the OBOR needs to deepen multi-stakeholder involvement, offering them efficient communication channels and giving more power for the stakeholders to steer the OBOR.

8.4.5 Public engagement and public debate

A permanent public platform can be helpful for engagement activities in line with the OBOR's two objectives. Meanwhile, public debate can hopefully have a moderating effect on the practice of balancing economic and ecological purposes.

The OBOR initiative has not fully addressed the importance of public engagement and public debate and has to some extent ignored it. From policy-making to planning processes, governments and technological elites have been the main actors. Eminent Chinese urban scholar Fulong Wu argues that planning has tended to be an un-political process since the Chinese Economic Reform, which means that citizens have had no vehicle for contributing to the conversation [61]. So far, public engagement for the OBOR has consisted of the Program for the Chinese Social Organizations, in which 90 nongovernmental organizations were recruited by the Chinese NGO Network for International Exchanges to promote the 'Belt and Road Initiative'. Although it is called an NGO programme, the Chinese government is the de facto organizer. For instance, the list of organizations that join the Societal Cooperation Network along the OBOR (入丝路沿线民间组织合作网络的组织名单) included numerous governmental organizations including 'All-China Youth Federation', 'All-China Women's Federation' and 'All-China Environment Federation' [62–64]. The absence of real public debate and monitoring public opinion of the OBOR initiative can mean that citizen voices go unheard as the government invests heavily in scientific and technological research and implementation, initiatives that may lead to the displacement of those very citizens. The stakes are very high for people to participate in the planning processes, and the Chinese authorities acknowledge it. According to the Urban and Rural Planning Law of the People's Republic of China, issued by the State Council in 2007, public engagement, specifically comments from the general public, help planning authorities to notice violations of the territory by the stakeholders and prevent illegal construction [65]. Provincial and local government have also started to prioritize public supervision of planning. But such platforms are currently temporary, as mentioned in the case of the Taipingwan area at Dalian Port. A continuous public platform should be established in the Dalian Port Cluster for the general public to share their views with policymakers and stakeholders and to better achieve the OBOR's goal of balancing economic and ecological objectives.

8.4.6 Port cluster: Enriching the methodology of the RRI

Our work here has enriched the work of other scholars, including the Dutch scholar in technological ethics, Jeroen van den Hoven [66]. It is also building on the work of Wim Ravesteijn, who has used the RRI to analyse port development projects in Dalian and Nansha ports [67, 68], but not to analyse spatial plans on a regional scale. This paper takes that step, contending that the development of Chinese ports in light of the OBOR initiative is related intimately to its neighbouring areas, thus, a port cluster. Assessing the Dalian Port Cluster can refine the use of the RRI. The multiple provincial governments in the port cluster extend the current mechanism of 'innovation governance and stakeholder involvement', which has not included the engagement of various governments from the same administrative level. Moreover, the horizontal communication channels between three provincial governments in the cluster also need to be transparent.

8.5 Conclusion

Examining the spatial impact of the OBOR initiative, this chapter first has posited the concept of the port cluster as a space in which to assess the implementation of the dual goals of the OBOR initiative: economic development and ecological sustainability. It argues that the port cluster is the appropriate planning scale for national spatial strategies of the OBOR initiative to coordinate economic development and ecological sustainability and that urban scholars, port logistic specialists, environmental and ecological experts and so on need to identify and conceptualize it. As is clear from the Dalian case, collaboration in the port cluster needs to be improved to achieve the OBOR's two objectives.

Second, the chapter has studied regional-scale planning in the Dalian Port Cluster, exploring barriers to planning and implementing change. Although the national government has formed the Dalian Port Cluster from various national economic plans and spatial strategies, the un-transparency and imbalance of economic development and ecological sustainability, the absence of a horizontal dialogue among the involved provinces and the absence of a vertical channel of communication between governments and the public all hinder the realization of the goals of the OBOR initiative.

Such shortcomings in the planning and executing of the OBOR initiative at the regional port cluster level may be due to the relatively short history of the OBOR initiative. Others are more systemic. Traditional top-down Chinese planning practices can be advantageous in realizing such a huge project, but also hinder transparent and open communication with governments at all scales and the general public.

Finally, the paper has used the RRI to figure out the different needs for improving regional-scale planning in order to imagine how to achieve the objectives of the OBOR initiative:

- The OBOR initiative needs more knowledge-based conferences or seminars for both scholars and the general public for notifications about probable risks.
- A mechanism for terminating negative projects is also required to reduce unexpected harm.
- Although public participation has been highlighted as an institutional priority in China, further political reform is needed for permanent and institutionalized public engagement.
- More political channels for local control over national planning concepts are needed to ensure the public's right to be included.
- The transparency of planning, a communication channel for hierarchical governments, channels for governments and industrial stakeholders and channels between governments and the general public are all required to

reduce unhealthy costs and achieve the main goal of sustainability in the OBOR initiative.

The RRI even generates further interest among urban scholars in exploring whether OBOR projects will gradually integrate the general public's comments into their spatial plans. Meanwhile, the concept of the port cluster enriches the methodology of RRI, making it more flexible for studying the built environment in port regions.

The Chinese government, provincial governments and other stakeholders involved in the OBOR initiative have been learning to avoid problems in planning, executing and supervision to better achieve the goals of the OBOR initiative. New concepts, methodologies and tools can further their work. This chapter is a contribution to that learning process.

Going forward, understanding how different Chinese governmental institutions translate the OBOR initiative into policies and plans within China will allow us to assess the role of local institutions and citizens in this continental-scale transformation; it will also help us understand the planned expansion of the Silk Road to ports and cities beyond China, where political and economic systems are often very different from those in China. The plan views the port of Hamburg, for example, as one of the main hubs for the New Silk Road. This north German port is already Europe's leading seaport for trade with China; among other dynamics, the high level of rail traffic (to transport containers) is possibly harming the environment [69]. Other European port cities, including Rotterdam, Antwerp and Piraeus, are also already experiencing the impact of Chinese policy [70]. For example, the nationally owned China Ocean Shipping (Group) Company has invested nearly half a billion euros in the port of Piraeus, making it the largest owner of the port. The OBOR initiative will extend Chinese influence beyond the port to reshape neighbouring port cities and their hinterlands. As it figures out how to balance economic and environmental concerns at home, perhaps it can also learn to extend that balance abroad.

To conclude, RRI is a promising and valuable methodology to assess the implementation of the OBOR initiative in the field of spatial planning. The OBOR initiative is still young, and its implementers are definitely in the midst of a learning process. Applying RRI to it now can help China assure that regional actors execute spatial plans in line with local needs and interests at the same time as they pursue economic and environmental sustainability wherever they build.

References

[1] National Development and Reform Commission; Ministry of Foreign Affairs; and Ministry of Commerce of the People's Republic of China;

Vision and Actions on Jointly Building Silk Road Economic Belt and 21st-Century Maritime Silk Road, 推动共建丝绸之路经济带和21世纪海上丝绸之路的愿景与行动, 2015.

[2] The One Belt, One Road Initiative (一带一路) is officially translated as the Belt and Road Initiative (BRI), and also known as the Silk Road Economic Belt (SREB), the New Silk Road, and colloquially as the new Maritime Silk Road; use of the OBOR initiative in this chapter reflects the most common and best-known acronym.

[3] Huang, Y. & Perlez, J., Behind China's $1 Trillion Plan to Shake Up the Economic Order, *The New York Times*, 13 May 2017.

[4] Hein, C., Port cities and urban waterfronts: How localized planning ignores water as a connector. *Wiley Interdisciplinary Reviews: Water*, 3(3), pp. 419–438, 2016.

[5] Hein, C., Oil spaces: The global petroleumscape in the Rotterdam/The Hague area. *Journal of Urban History,* 44 (5), pp. 887–929, 2018.

[6] Mullany, G., Huge Oil Spill Spreads in East China Sea, Stirring Environmental Fears, *The New York Times*, 15 January 2018.

[7] Xinhua Net, China Reveals Risk Map of Marine Disasters. http://www.xinhuanet.com/english/2018-05/12/c_137174555.htm. Accessed on: 21 May 2021.

[8] China Opens Free Trade Zone in Shanghai; *Deutsche Welle*, https://www.dw.com/en/china-opens-free-trade-zone-in-shanghai/a-17121642. Accessed on: 29 September 2013.

[9] Plan for Revitalizing the Old Industrial Bases in Northeast China, 东北地区振兴规划. Beijing: National Development and Reform Commission, 2006.

[10] The State Council's Decision on Implementing the Scientific Outlook on Development and Strengthening Environmental Protection, 国务院关于落实科学发展观加强环境保护的决定. Beijing: The State Council, 3 December 2005.

[11] Sutcliffe, H., A report on Responsible Research & Innovation, Director, MATTER (On the basis of material provided by the Services of the European Commission. Prepared for DG Research and Innovation, European Commission). https://ec.europa.eu/research/science-society/document_library/pdf_06/rri-report-hilary-sutcliffe_en.pdf. Accessed on: 21 May 2021.

[12] Yang, P. & Han, B., Responsible research and innovation and its implications for China. *China and World Economy,* 25(6), pp. 120–138, 2017.

[13] Sternberg, T., Ahearn, A., & McConnell, F., Central Asian 'characteristics' on China's New Silk Road: The role of landscape and the politics of infrastructure. *Land,* 6(3), pp. 55–73, 2017.

[14] Tracy, E.F., Shvarts, E., Simonov, E. & Babenko, M., China's new Eurasian ambitions: The environmental risks of the Silk Road Economic Belt. *Eurasian Geography and Economics,* 58(1), pp. 56–88, 2017.

[15] Fardella, E. & Prodi, G., The Belt and Road initiative impact on Europe: An Italian perspective. *China & World Economy,* 25(5), pp. 125–138, 2017.

[16] Shariatinia, M. & Azizi, H., Iran–China cooperation in the Silk Road Economic Belt: From strategic understanding to operational understanding. *China and World Economy,* 25(5), pp. 46–61, 2017.

[17] Schinas, O. & Westarp, A.G., Assessing the impact of the Maritime Silk Road. *Journal of Ocean Engineering and Science,* 2(3), pp. 186–195, 2017.

[18] Chhetri, P., Nkhoma, M., Peszynski, K., Chhetri, A. & Lee, P.T.-W., Global logistics city concept: A cluster-led strategy under the Belt and Road initiative. *Maritime Policy & Management*, 45(3), pp. 319–335, 2017.

[19] Lee, P.T.-W., Hu, Z.-H., Lee, S.-J., Choi, K.-S. & Shin, S.-H., Research trends and agenda on the Belt and Road (B&R) initiative with a focus on maritime transport. *Maritime Policy & Management*, 45(3), pp. 282–300, 2018.

[20] de Langen, P.W. & Visser, E.-J., Collective action regimes in seaport clusters: The case of the Lower Mississippi Port Cluster, *Journal of Transport Geography,* 13(2), pp. 173–186, 2005.

[21] Monios, J. & Wang, Y., Spatial and institutional characteristics of inland port development in China. *Geo Journal,* 78(5), pp. 897–913, 2013.

[22] Wu, F. & Phelps, N.A., From suburbia to post-suburbia in China? Aspects of the transformation of the Beijing and Shanghai Global City Regions. *Built Environment,* 34(4), pp. 464–481, 2008.

[23] Lin, Y., Urban regeneration in China: Policy, development, and issues. *Local Economy: The Journal of the Local Economy Policy Unit*, 26(5), pp. 337–347, 2011.

[24] Zhang, L. & Zhao, S. X., City branding and the Olympic effect: A case study of Beijing. *Cities*, 26(5), pp. 245–254, 2009.

[25] Ye, L. & Björner, E., Linking city branding to multi-level urban governance in Chinese mega-cities: A case study of Guangzhou. *Cities*, 80, pp. 29–37.

[26] Bie, J., Jong, M. de & Derudder, B., Greater Pearl River Delta: Historical evolution towards a global city-region. *Journal of Urban Technology*, 22(2), pp. 103–123, 2015.

[27] Xiong, L., Yu, C., de Jong, W.M., Wang, F. & Cheng, B., Economic transformation in the Beijing-Tianjin-Hebei region: Is it undergoing the environmental Kuznets Curve? *Sustainability*, 9(5), pp. 869–869, 2017.

[28] de Jong, M., Chen, Y., Joss, S., Lu, H., Zhao, M., Yang, Q. & Zhang, C., Explaining city branding practices in China's three mega-city regions: The role of ecological modernization. *Journal of Cleaner Production,* 179, pp. 527–543, 2018.

[29] Han, M., de Jong, M., Cui, Z., Xu, L., Lu, H. & Sun, B., City branding in China's northeastern region: How do cities reposition themselves when facing industrial decline and ecological modernization? *Sustainability,* 10(1), article 102, 2018.

[30] Ministry of Ecology and Environmental, Ministry of Foreign Affairs, National Development and Reform Commission, Ministry of Commerce, Guidance on Promoting the Construction of the Green 'Belt and Road', 关于推进绿色"一带一路"建设的指导意见, 2017.

[31] Ministry of Ecology and Environmental, The Belt and Road Ecological and Environmental Cooperation Plan, 2017.

[32] National People's Congress, Outline of the 13th Five-Year Plan for the National Economic and Social Development of the People's Republic of China, 中华人民共和国国民经济和社会发展第十三个五年规划纲要, 2016.

[33] The State Council, Territory Development Plan, 全国主体功能区规, 2010.

[34] The State Council, Marine Development Plan, 全国海洋主体功能区规划 2015.

[35] Sun, X., When Will the Environmental Protection Industry Become the Backbone of the Domestic Economy, 环保产业何时能成经济"台柱子环保产业何时能成经济"台柱子, *People's Daily*, 人民日报. http://env.people.com.cn/n1/2017/0805/c1010-29451233.html. Accessed on: 24 June 2021.

[36] The National Development and Reform Commission and the State Oceanic Administration, Vision for Maritime Cooperation under the Belt and Road Initiative, '一带一路'建设海上合作设想. https://www.yidaiyilu.gov.cn/wcm.files/upload/CMSydylgw/201706/201706200153032.pdf. Accessed on: 20 June 2017.

[37] National Development and Reform Commission, The Thirteenth Five-Year Plan for Revitalizing the Old Industrial Bases in Northeast

China, 十三五东北地区振兴规划. https://www.gov.cn/xinwen/2016-12/20/5150168/files/7779d2ddc7d24744ac036acaf6a89037.pdf. Accessed on: 24 June 2021.

[38] Xie, L., Yang, Z., Cai, J., Cheng, Z., Wen, T and Song, T., Harbin: A rust belt city revival from its strategic position. *Cities*, 58, pp. 26–38, 2016.

[39] National Energy Administration & National Development and Reform Commission, Medium- and Long-Term Plan of Oil and Gas Pipeline Network, 中长期油气管网规. https://www.ndrc.gov.cn/xxgk/zcfb/ghwb/201707/W020190905497932558033.pdf. Accessed on: 24 June 2021.

[40] Pardali, A., Kounoupas, E. & Lainos, I., Can clusters be bi-polar? Exploring the case of the Piraeus Port-maritime cluster, *Maritime Policy & Management*, 43(6), pp. 706–19, 2016.

[41] People's Government of Liaoning, The Thirteen Five-Year Plan for the Economic and Social Development of Liaoning Province, 辽宁省国民经济和社会发展第十三个五年规划纲要. http://www.cfen.com.cn/sjpd/sj/201704/P020170420584090400575.pdf. Accessed on: 23 March 2016.

[42] National Marine Environment Monitoring Centre, The Plan of the Marine Development of Liaoning Province, 辽宁省海洋主体功能区规划; http://www.ln.gov.cn/zwgkx/ghjh/201911/t20191129_3645498.html. Accessed on: 24 June 2021.

[43] Bureau of Urban Planning in Dalian, The 2017 Revised Comprehensive Urban Plan of Dalian (2001–2020), 大连市城市总体规划 (2001-2020年) (2017年修订). https://zrzy.dl.gov.cn/art/2017/4/27/art_3082_504681.html. Accessed on: 24 June 2021.

[44] Dalian Port Group Corporation Limited, Plan towards the Green Ecological Port, 绿色生态港口发展规划. https://wenku.baidu.com/view/fab556c42af90242a995e5a0.html. Accessed on: 24 June 2021.

[45] Fei, Y., The Dalian Port Has Set Up the 'Protection' of the Oil Storage Tanks from Lightning, 我港搭起油罐区防雷"保护伞. http://www.portdalian.com.cn. Accessed on: 29 July 2016.

[46] Sina Comprehensive, Petrochina (Dalian) Petrochemical Has Frequent Fires: It Is the Seventh Accident since 2010, 中石油大连石化火灾频发:2010年至今已是第七次事故. http://finance.sina.com.cn/chanjing/gsnews/2017-08-18/doc-ifykcypp8898450.shtml. Accessed on: 18 August 2017.

[47] Diao, F., Taipingwan Port (Dalian) Has Been Fined 10 Million for Illegal Reclamation, and Local Subsidies Are All Paid Back, 太平湾港区违法填海被罚千万当地又以补贴全返.

http://m.thepaper.cn/newsDetail_forward_1747484. Accessed on: 31 July 2017.

[48] Thirtieth Meeting of the Standing Committee of the Tenth National People's Congress of the People's Republic of China, The Urban and Rural Planning Code of the People's Republic of China, 中华人民共和国城乡规划法. http://www.npc.gov.cn/npc/xinwen/lfgz/zxfl/2007-10/28/content_373842.htm. Accessed on: 28 October 2007.

[49] Dalian Port Company, Publicities of the Environmental Impact Assessment of the Dalian Port Taipingwan Area Master Plan (First Time), 大连港太平湾港区总体规划环境影响评价公示（第一次）. http://www.gkj.dl.gov.cn/behavior/article.php?COLLCC=247333979&id=17090815231754&zid=81. Accessed on: 8 September 2017.

[50] Dalian Port Company, Publicities of the Environmental Impact Assessment of the Dalian Port Taipingwan Area Master Plan (Second Time), 大连港太平湾港区总体规划环境影响评价公示（第二次）. http://www.gkj.dl.gov.cn/behavior/article.php?COLLCC=247333979&id=17122912065919&zid=81. Accessed on: 29 December 2017.

[51] Simakova, E. & Coenen, C., Visions, hype, and expectations: A place for responsibility. In: Bessant, J., Heintz, M. & Owen, R. (eds), *Responsible Innovation: Managing the Responsible Emergence of Science and Innovation in Society*. Chichester: Wiley, pp. 241–268. 2013.

[52] Directorate-General for Research and Innovation, Directorate B – European Research Area, and Unit B.6 – Ethics and Gender European Commission. Options for Strengthening Responsible Research and Innovation: Report of the Expert Group on the State of Art in Europe on Responsible Research and Innovation. https://op.europa.eu/en/publication-detail/-/publication/1e6ada76-a9f7-48f0-aa86-4fb9b16dd10c. Accessed on: 19 May 2021.

[53] European Commission, Responsible Research and Innovation: Europe's Ability to Respond to Societal Challenges. https://op.europa.eu/en/publication-detail/-/publication/bb29bbce-34b9-4da3-b67d-c9f717ce7c58. Accessed on 19 May 2021.

[54] von Schomberg, R., Prospects for technology assessment in a framework of responsible research and innovation. In: Dusseldorp M., Beecroft, R., (eds) *Technikfolgen abschätzen lehren*. Wiesbaden: VS Verlag für Sozialwissenschaften, pp. 39–61, 2012.

[55] von Schomberg, R., A vision of responsible research and innovation. In: Bessant, J., Heintz, M. & Owen, R. (eds), *Responsible Innovation: Managing the Responsible Emergence of Science and Innovation in Society*. Chichester: Wiley, pp. 51–74, 2013.

[56] National Natural Science Foundation of China. http://www.nsfc.gov.cn/english/site_1/index.html. Accessed on: 19 May 2021.

[57] National Natural Science Foundation of China, Support Scientific Personnel Cooperation to Create 'Belt and Road' Future – Science Fund 'Belt and Road' International Symposium Successfully Held, 支持科学人才合作共创"一带一路"未来——科学基金"一带一路"国际研讨会成功举办. http://www.nsfc.gov.cn/publish/portal0/tab445/info69637.htm. Accessed on: 24 June 2021.

[58] National Science Foundation of China, Cooperation Action Plan of the National Science Foundation of China to Build One Belt One Road Initiative, 中国科学院共建"一带一路"国际科技合作行动方案, *Bulletin of Chinese Academy of Sciences*, 32(Z2), pp. 8–13, 2017.

[59] National Development and Reform Commission & The State Oceanic Administration, Vision for Maritime Cooperation under the Belt and Road Initiative; 一带一路'建设海上合作设想. https://www.yidaiyilu.gov.cn/wcm.files/upload/CMSydylgw/201706/201706200153032.pdf. Accessed on: 19 May 2021.

[60] Zhong, W., Liu, M. & Zhang, G., Written in the New Constitution: Promoting the Fate of Human Community is a Continued Positive for the OBOR Initiative, 人类命运共同体"写入宪法持续利好"一带一路. https://www.yidaiyilu.gov.cn/xwzx/gnxw/50166.htm. Accessed on: 14 March 2018.

[61] Wu, F., *Planning for Growth: Urban and Regional Planning in China*. New York: Routledge, The RTPI Library Series, 2015.

[62] China NGO Network for International Exchanges, Program for the Chinese Social Organizations to Promote the 'Belt and Road Initiative', 中国社会组织推动"一带一路"民心相通行动计划. https://www.yidaiyilu.gov.cn/zchj/jggg/36736.htm. Accessed on: 27 November 2017.

[63] China NGO Network for International Exchanges, List of Organizations which Have Joined the Societal Cooperation Network along the OBOR, 加入丝路沿线民间组织合作网络的组织名单. https://www.yidaiyilu.gov.cn/zchj/jggg/36736.htm. Accessed on: 27 November 2017.

[64] China NGO Network for International Exchanges, List of Organizations which Have Joined the Action Plan of the Chinese Social Organizations to Promote the 'One Belt, One Road' (2017–2020), 加入中国社会组织推动"—带—路"民心相通行动计划(2017–2020)的组织名单. https://www.yidaiyilu.gov.cn/zchj/jggg/36736.htm. Accessed on: 27 November 2017.

[65] State Council, The Urban and Rural Planning Code of the People's Republic of China, 中华人民共和国城乡规划法. http://www.gov.cn/flfg/2007-10/28/content_788494.htm. Accessed on: 28 October 2007.

[66] van den Hoven, J., Wildschut, E. & Bakker, E., Responsible innovation at port cities. *Bollettino del Dipartimento di Conservazione dei Beni Architettonici ed Ambientali,* 12(1), pp. 765–770, 2012.

[67] Ravesteijn, W., He, J. & Chen, C., Responsible innovation and stakeholder management in infrastructures: The Nansha Port Railway Project. *Ocean & Coastal Management*, 100, pp. 1–9, 2014.

[68] Ravesteijn, W., Liu, Y. & Yan, P., Responsible innovation in port development: The Rotterdam Maasvlakte 2 and the Dalian Dayao Bay extension projects. *Water Science and Technology: A Journal of the International Association on Water Pollution Research*, 72(5), pp. 665–677, 2015.

[69] Port of Hamburg, Hamburg One of the Most Vital Hubs for the Silk Road, https://www.hafen-hamburg.de/en/news/hamburg-one-of-the-most-vital-hubs-for-the-silk-road—35326. Accessed on: 3 July 2017

[70] Horowitz, J. & Alderman, L., Chastised by E.U., a Resentful Greece Embraces China's Cash and Interests. *The New York Times*, https://www.nytimes.com/2017/08/26/world/europe/greece-china-piraeus-alexis-tsipras.html. Accessed on: 26 August 2017.

CHAPTER 9

Addressing value conflicts for the new Maritime Silk Road: Sustainable and responsible cost–benefit analysis in the Port of Shanghai

Lili Song

School of Management, Harbin Normal University, China

Abstract

Benefiting from the Belt and Road initiative, the Port of Shanghai, as the world's largest container port, is striving for sustainable and responsible development in the future, which is not only significant for local residents but also, potentially, for foreign ports. In this process, how to balance economic and other social benefits, with regard to environmental, health and ecosystem conservation, is a challenge. This amounts to dealing with conflicting values from various stakeholders in developing the port. Along with the Maritime Silk Road initiative, the internal and external circumstances the Port of Shanghai is facing are much more complicated than before. Therefore, in this chapter, strategies and actions carried out by the Port of Shanghai will be discussed, while an improved cost–benefit analysis is proposed to deal with value conflicts of different stakeholders.

Keywords: Responsible Innovation, value conflicts, cost–benefit analysis, Maritime Silk Road, Port of Shanghai

9.1 Introduction

In 2013, the Chinese President, Xi Jinping, announced the One Belt and One Road (OBOR) initiative, including the 21st Century Maritime Silk Road, aiming to promote common prosperity of the regions along the route. The new Maritime Silk Road covers almost all major container ports in the world and will constitute a crucial driving force in the development of global maritime trade and the world economy as a whole [1]. Ports will serve as a pivot in constructing the 21st Century Maritime Silk Road. The Port of Shanghai, as the world largest container port, plays an important role in the new Maritime Silk Road. The Port of Shanghai is devoted to being sustainable and responsible through various strategies and actions [2]. Consequently, it will not only be an engine of economic growth for the countries along the Silk Road, but also a good example for foreign ports along the route.

9.2 The Port of Shanghai and the OBOR: Chances, strategies and responsibilities

9.2.1 Overview of the port

The Port of Shanghai is situated in the middle of the 18,000 km Chinese coastline, where the Yangtze River, known as 'the Golden Waterway', flows into the sea. It is the meeting point in the T-shaped waterway network composed of the east–west Yangtze River and the south–north coastline, and it is also China's largest comprehensive port and one of the country's most important gateways for foreign trade [2]. The Port of Shanghai has extensive access to different modes of transportation. It faces the northern and southern coastal seas of China and the oceans of the world. It is linked to the Yangtze River and the inland waters of the Yangtze River Valley region, with connections to the provinces of Jiangsu, Zhejiang, Anhui and so on. Expressways and state-level highways lead from the Port of Shanghai to the national highway network to all regions of the country. Consequently, the Port of Shanghai enjoys a favourable geographical location, favourable natural conditions, extensive economically developed hinterland areas and complete infrastructure and facilities for distribution inland.

Driven by the New Silk Road initiative, the Port of Shanghai is striving to maintain a rapid and healthy development involved in containerization by implementing the Yangtze River strategy, Northeast Asia strategy and Internationalization strategy, concentrating on the four pillar activities of cargo handling, port logistics, port services and port commerce to establish and consolidate its status as the pivotal port of international shipping and to become the world's leading terminal operator and provider of port logistics services.

9.2.2 Yangtze River strategy

As the direct hinterland of the port of Shanghai, the Yangtze River Valley is the base of the port and a critical area for its sustainable growth. It is the cornerstone of the development of the Port of Shanghai now and in the future to maintain and increase its share of the containerization market. The key to the Yangtze River strategy is to lead the direction of the flow of goods, enhance the port's influence over the river basin and realize the Sustainable Development of the Port of Shanghai. As the scene of the port's direct cargo traffic, the Yangtze River Valley is the foundation of the development of the home port and crucial to its future volume growth. It is vital for the Port of Shanghai to maintain and increase its share in the Yangtze River Valley container market [2].

Figure 9.1: Yangshan deep-water port. (Source: Wikimedia Commons, 2018.)

9.2.3 Northeast Asia strategy

As an indirect hinterland of the Port of Shanghai, Northeast Asia provides room for the growth of the port and it is the key area for Shanghai to be an international shipping centre. For the future, it is a critical step in the rapid development of the Port of Shanghai to ensure its foothold in the container trans-shipment market in Northeast Asia, based on the Yangshan deep-water port, Shanghai port's outpost in the ocean (see Figure 9.1). Covering the area of the port's indirect cargo traffic, the Northeast Asia strategy is a driver for the Port of Shanghai to take a leap forward to enhance container levels in Northeast Asia [2].

9.2.4 Internationalization strategy

Exploring overseas markets is an efficient means for materializing an internationalization strategy. The Port of Shanghai is likely to create value for shareholders through the acquisition of foreign terminals and engaging in businesses relevant to the port. This strategy, aimed at forming a transnational network that covers both domestic and overseas markets with the know-how and expertise in port management and operation and innovative services of financing in port business, will help the Port of Shanghai become an international terminal operator rendering service to customers along the whole supply chain [2]. The Port of Shanghai, benefiting from the Maritime Silk Road initiative, has more opportunities, which means more sustainable social responsibilities, not only for local residents but also for foreign ports along the new Maritime Silk Road.

9.3 Sustainable Development and Responsible Innovation: The challenge of value conflicts

Today, world ports have to deal with the rising significance of sustainability issues caused by climate change (causing sea level rise and changing river discharges) and air and water pollution as well as growing pressures on scarce resources, besides other rapid urbanization issues (especially in fast-developing and newly emerging economies). What is more, these issues have to be addressed in a situation of expanding decision- and policymaking arenas, with increased involvement of (critical) stakeholders, especially citizens. In addition, trends in global trade and environmental regulations are highly uncertain. Therefore, the goal of port development should shift from one-sided economic growth to comprehensive sustainable and responsible development. This necessitates dealing with diverging, sometimes competing or even conflicting values, caused by the double aim of port development – economic growth and prosperity improvement – in order to reconcile and balance different values and finding innovations that allow us to satisfy more than one value at the same time [3–5].

In this respect, Responsible Innovation refers to:

> ways of proceeding in research and innovation that allow those who initiate and are involved in these processes at an early stage (A) to obtain relevant knowledge on the consequences of the outcomes of their actions and on the range of options open to them and (B) to effectively evaluate both outcomes and options in terms of ethical values (including, but not limited to well-being, justice, equality, privacy, autonomy, safety, security, sustainability, accountability, democracy and efficiency) and (C) to use these considerations (under A and B) as functional requirements for design and development of new research, products and services [6].

Responsible Innovation could help to deal with these different and conflicting values.

To implement Responsible Innovation, a cost–benefit analysis (CBA) could be applied, which is always used to assess the net profits of public projects. In addition to the economic consequences of port development, social effects such as the consequences for the environment, health and so on are much more important for society. Therefore, a normal CBA could be improved to a social

CBA, in which social aspects, such as people and the planet, instead of just profit, are included in the evaluation process.

9.4 Sustainable/responsible development of the Port of Shanghai

9.4.1 Ambitions and actions

Since 2008, the Port of Shanghai publishes not only a yearly financial report but also a yearly Sustainable Development Report. The latter report reflects changes occurring in the public domain after the development of the Yangshan deep-water port project in 2002. As indicated previously, Chinese society is changing fast, and citizen values are increasingly coming to the fore, especially in worldly cities like Shanghai. The port recognizes this, increasingly giving attention to ecological issues, as in the Sustainable Development Report [7], in which the mission of the Port of Shanghai is described as follows:

> We continuously act on the basis of social responsibility, the core of which is realizing sustainable development, which can be explained as that we will take responsibility for the stakeholders and the environment in pursuing economic profits; and eventually, for realizing harmony between the development of business and society, and between the development of Shanghai port and environmental sustainability.

Actions have been taken to achieve the Sustainable Development of the Port of Shanghai, including the construction of a smart port, green port (or eco-port), technological port and an efficient port, as shown in Figure 9.2. The smart port is based on the Internet of Things and Big Data, aiming to realize automated terminals. The green port or eco-port focuses on saving resources and being environmentally friendly and actively fulfilling social responsibilities. Various technologies and measurements are applied to save resources and energy to protect the ecological environment and combat climate change. The construction of a technological and efficient port mainly relies on technical, management and service innovation, such as an intelligent management system, a tally wireless control system and an unmanned automatic container yard, leading to the optimization of the allocation of resources.

9.4.2 Cost–benefit analysis

A CBA is the best-known and most applied method in evaluating public sector projects. The purpose of a CBA is to provide a consistent procedure for assessing the net economic benefits of proposed policies or projects to a defined group. To the extent possible, a CBA should identify and quantify all-important monetary and non-monetary costs and benefits. Therefore, the scope of a

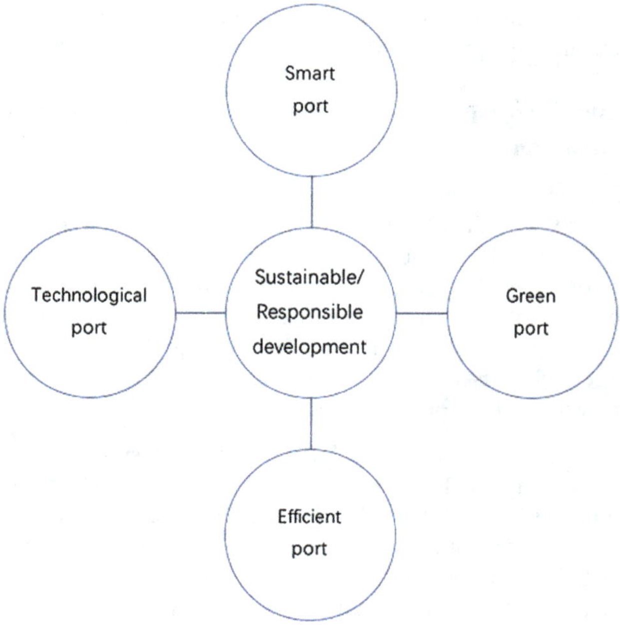

Figure 9.2: Sustainable/responsible development of the Port of Shanghai.

cost–benefit analysis is far broader and much more inclusive than a financial feasibility analysis. However, it is much more difficult because many important costs and benefits cannot be fully captured and considerable detective work is required to uncover the value the public holds for non-monetary goods and services [8–12].

9.4.3 Drawbacks

Due to the uncertainty in determining non-monetary aspects, the application of a CBA has provoked a lot of criticism, which could be summarized into five main points. First, Scanlon pointed out that the satisfaction of people's 'manifest' preferences is not an adequate index of well-being because there are circumstances conceivable in which these preferences might be satisfied, even though the individuals' true interests were far from being served [13]. This seems to suggest that people cannot be relied on to follow and advance their own interests and that somebody else – another actor – must declare certain goods and conditions as important for a good life; for example, someone will have to stand up for the protection of nature and endangered species when building a new and economically attractive port in a wildlife area [14].

Second, Niemeyer and Spash as well as Heinzerling and Ackerman argued that the common methods of valuing environmental impacts in a CBA cannot possibly result in a true picture of the real values people attach to such goods as nature, landscapes or a clean environment; therefore, market analogy valuation methods for non-monetary goods are inherently flawed [15, 16].

A third criticism is that a CBA ignores equity issues [16] since it does not pose the question of who suffers as a result of, for example, environmental problems and, therefore, threatens to reinforce existing patterns of economic and social development.

The fourth point of dispute is anxiety about incomplete or incorrect information being supplied to decision-makers. Mackie and Preston argued that impacts that are difficult to value in monetary terms tend to be excluded. They particularly emphasized the real danger of unbalanced treatment of monetary versus non-monetary impacts [17]. The risk here is that the non-monetary impacts will be represented less dominantly in the conclusion compared with the priced impacts.

The last dispute is that a CBA fails to produce greater objectivity and transparency promised by its proponents. It seems that the highly complex, resource-intensive and expert-driven nature of a CBA makes it extremely difficult for the public to understand and participate in the process.

9.4.4 Improvement of a CBA for sustainable/responsible port development

Although some disadvantages have been addressed by researchers, a CBA is still widely used to assess the impacts of public projects or policies on society. Traditional CBAs could be improved by extending more social elements and stakeholders in the analysis to broaden the scope of the CBA. Regarding the Port of Shanghai, in the context of the Maritime Silk Road initiative, value for stakeholders as well as future generations in other countries along the route plays a significant role in addition to cost efficiency. Therefore, an improved CBA for the Port of Shanghai – to be applied to new projects and developments in general or, as suggested here, to the whole of port operations – must be based on a broader set of viewpoints, so that the Port of Shanghai can be an example for foreign ports along the new Maritime Silk Road, in addition to serving local residents. The costs and benefits of the Port of Shanghai are listed in Table 9.1.

In this extended CBA, benefits for foreign countries along the route are included. Some of their seaports are rebuilt or extended in response to the Maritime Silk Road initiative and, consequently, these new ports will be able to participate in more international trade activities over the world, leading to greater economic growth and more jobs for the local economy. Furthermore, the more international trade activities, the more use of production factors such as capital, labour and technology, which will result in technological innovation, especially the application of efficiency-oriented techniques. Finally, increasing demands on ports will absorb more investment and other transport infrastructures like airports, highways, railways and inland waterways.

Table 9.1: Element description of the social cost–benefit analysis of the Port of Shanghai.

Elements	Description
Costs	
Investment	Incremental investment costs for planning, port infrastructure and offsite investments (e.g. bridges, road connections and overpasses).
Operation and maintenance	Regular expenditure for daily operations (e.g. payments for port workers).
Health and safety	Environmental safety (e.g. emissions and industrial accidents); nautical safety; structural safety; health risks (number of injuries and deaths).
Ecological compensation	Biological diversity in the port area.
Environment sustainability	Reducing emissions and decreasing the use of raw materials and energy.
Benefits for local residents	
Transportation	Transportation cost savings with the port.
Energy savings	Benefits from energy savings (e.g. water, fuel and electricity).
Landscape recreation	Environmental benefits of the port after mitigation.
Benefits for foreign countries along the new Maritime Silk Road	
International trade	Increase of import and export via cooperation with the Port of Shanghai
Technology innovation	Application of efficiency-oriented techniques.
Employment	Increased supply of jobs.
Investment in transport infrastructures	Increasing demands for the port, which will absorb more investments into local ports and other transport infrastructures.

9.5 Conclusion

Many countries along the Silk Road could benefit from the Belt and Road initiative, especially countries that are located along the seacoast since seaports not only have strong economic influences on port cities but also on the hinterlands they are serving. Therefore, involved governments and port authorities increasingly invest in ports, aiming to increase the capacities and competitiveness of the ports. However, in the process of constructing and operating ports, environmental issues and, in general, value conflicts, emerge. Being sustainable and responsible could be a solution for ports to cope with environmental problems and conflicts.

The Port of Shanghai is acting to be sustainable and responsible in order to be consistent with the Maritime Silk Road initiative, but the port authority is facing serious and apparently conflicting challenges. On one hand, increasing international trade and continuous technological development necessitate further development of the Port of Shanghai, expanding capacities. On the other hand, social and environmental concerns complicate these developments. Therefore, in this chapter, an improved cost–benefit analysis is proposed and recommended to deal with the port's double aim. Stakeholders from abroad along the route are also involved in the CBA to make the analysis more inclusive in order to achieve sustainable and responsible development, not only for local residents but also for the countries and ports along the new Maritime Silk Road.

References

[1] Ministry of Commerce, People's Republic of China. http://www.mofcom.gov.cn/article/ae/ai/201503/20150300928878.shtml. Accessed on: 21 May 2021.

[2] Port of Shanghai. https://web.archive.org/web/20210301122732/http://en.portshanghai.com.cn/. Accessed on: 2 July 2021.

[3] Song, L. & Ravesteijn, W., Responsible port innovation in China: The case of the Yangshan Port Extension Project. *International Journal of Critical Infrastructures*, 11(4), pp. 297–315, 2015.

[4] Taneja, P., Ros, R., Vellinga, T. & Schuylenburg, M. van, Flexibility in port infrastructures. The 8th International Conference on Coastal and Port Engineering in Developing Countries, Chennai, India, 2012.

[5] Taneja, P., Ligteringen, H. & Walker, W.E., Flexibility in port planning and design. *European Journal of Transport and Infrastructure Research*, 12(1), pp. 65–86, 2012.

[6] Hoven J., van den & Jacob, K., Options for Strengthening Responsible Research and Innovation. Report of the Expert Group on the State of

Art in Europe on Responsible Research and Innovation, European Union, Brussels, 2013.

[7] Port of Shanghai, Sustainable Development Report of Shanghai Port. Shanghai, China, 2008–2017.

[8] Rajesh, G., Saravanan, R., Ravichandran, V., Parameswara Pandian, S. & Ramani Sujatha, E., Cost benefit analysis on different configurations of berthing structure. *IOP Conference Series: Earth and Environmental Science*, 80(1), 012048, 2017.

[9] Reyns, N., Casaer, J. & Smet, L.D., Cost-benefit analysis for invasive species control: The case of greater Canada goose *Branta canadensis* in Flanders (northern Belgium). *Peerj*, 6(e4283), 2018.

[10] Bonnieux, F., Goffe, P.L., Dabbert, S., Dubgaard, A., Slangen, L. & Whitby, M., Cost-benefit analysis of landscape restoration: A case-study in western France. In: Dabbert, S., Dubgaard, A., Slangen, L. & Whitby, M. (eds), *Economics of Wildlife and Landscape Conservation*. Wallingford: CAB-International, pp. 85–96, 2017.

[11] Atkinson, G., Groom, B., Hanley, N. & Mourato, S., Environmental valuation and benefit-cost analysis in U.K. policy. *LSE Research Online Documents on Economics*, pp. 1–23, 2018.

[12] Jin, Y., Andersson, H. & Zhang, S., China's cap on coal and the efficiency of local interventions: A benefit-cost analysis of phasing out coal in power plants and in households in Beijing. *Journal of Benefit-Cost Analysis*, 8, pp. 1–40, 2017.

[13] Scanlon, T.M., The moral basis of interpersonal comparisons. In: Elster, J. & Roemer, J.E., *Interpersonal Comparisons of Well-Being*. Cambridge: Cambridge University Press, pp. 17–44, 1991.

[14] Annema, J.A., Koopmans, C. & Wee, B. van, Evaluating transport infrastructure investments: The Dutch experience with a standardized approach. *Transport Reviews* 27(2), pp. 125–150, 2007.

[15] Niemeyer, S.C. & Spash, L., Environmental valuation analysis, public deliberation, and their pragmatic syntheses: A critical appraisal. *Environment and Planning C: Government and Policy*, 19, pp. 567–585, 2001.

[16] Heinzerling, L. & Ackerman, F., *Pricing the Priceless: Cost–Benefit Analysis of Environmental Protection*. Washington, DC: Georgetown Environmental Law and Policy Institute, Georgetown University Law Center, 2002.

[17] Mackie, P. & Preston, J., Twenty-one sources of error and bias in transport appraisal. *Transport Policy*, 5(1), pp. 1–7, 1998.

CHAPTER 10

Construction in harmony: The case of Yancheng Port

Jinxin Zhang[1,2], Wim Ravesteijn[1,3] & Zhijie Li[1,4]

[1]Department of Values, Technology and Innovation, Delft University of Technology, The Netherlands
[2]College of Water Conservancy and Hydropower Engineering, Hohai University, China
[3]Beijing Information Science and Technology University, China
[4]School of Management, Harbin Institute of Technology, China

Abstract

In view of the ambitious Maritime Silk Road programme of the Chinese government, this chapter describes and specifies a novel approach to deal with value conflicts in port construction projects: Responsible Port Innovation (RPI). Stakeholder analysis and impact assessments can be used to explore and inventory the potential value tensions that might arise, while several methods are available for addressing these issues, including Cost–benefit analysis (CBA) and process management methods. The current extension project of Yancheng Port is used as a case study. In tackling value oppositions, CBA and the so-called Harmony model will be used. Both methods are compared and assessed to find out which is the most suitable for port projects in China. We will argue that RPI can lead to better port projects, especially in combination with a special, localized approach (in this case the Harmony model) for dealing with value conflicts.

Keywords: Responsible Innovation, Yancheng Port, the Maritime Silk Road, Harmony model

10.1 Introduction

The New Silk Road project, also known as 'One Belt, One Road' (OBOR), is an ambitious development plan and framework, proposed by Xi Jinping, the president of China since 2013, which focuses on promoting Sustainable Development and peace through improving the connectivity of Asia, Europe and Africa. It consists of two main components, the land-based 'Silk Road Economic Belt' and the oceangoing 'Maritime Silk Road'. According to the planning report of the Maritime Silk Road, in the next two decades, 15 large-scale port projects will be executed in seven provinces along the coast of China. The 15 ports, from the north to south, cover almost the whole of China's coastal areas, involving a population of up to 126 million people, equivalent to about a quarter of the EU's total population [1]. More than €2 billion will be invested in the construction of these ports, which will bring thousands of jobs to local people. However, this ambitious plan will not only stimulate economic development but also bring a series of environmental issues and value tensions.

The twofold New Silk Road is facing great challenges. From the economic side, some countries in the Eurasian region have a weak economic foundation, backwards infrastructure and a heavy debt burden. Politically, the dispute over the South China Sea is still pending. Political instability in some countries and regions is seriously affecting the continuity of economic policies. From a cultural perspective, the New Silk Road covers a wide range of different cultures, which may lead to violent conflicts of a 'civilizational' nature. Meanwhile, environmentally, the geographical features of the Eurasian continent, including an unusually dry climate, extremely scarce rainfall and large deserts, are hard to cope with in constructing the land route, while the sea route will cause extra pollution and biodiversity reduction. In general, the fragility of the ecological environment affects the carrying capacity of human and economic activities, which are testing the existing ecological boundaries in supporting the huge scale of development. These factors together determine that the New Silk Road is a promising but risky undertaking in infrastructure interconnection, long-term investment and other development. If these environmental problems and the value tensions, in general, cannot be addressed adequately, this series of port and other infrastructure projects may not be carried out smoothly, and even lead to the failure of the New Silk Road project.

As can be concluded from previous experience and scholarship, port and other infrastructure projects directly relate to local residents' interests and values. Consequently, these kinds of projects should be tackled with consideration for all stakeholders and their input, which requires new approaches. Responsible Innovation (RI) is such an approach. It aims at designing, developing and managing innovation trajectories, including all stakeholders, analysing their goals and interests, and preventing and solving value tensions between them [2]. RI is becoming increasingly important and relevant in large port extension and other complex technological projects in Europe, where it first cropped up [2], but it is not common in other parts of the world, especially not in developing countries, including China.

In this chapter, RI will be introduced and applied to port development, illustrated and explained by the presently planned Yancheng Port Extension Project in Jiangsu province in the east of China (not far from the Port of

Shanghai). We will focus on value conflicts between and among developers and stakeholders and discuss several methods to address these, especially Cost–benefit analysis (CBA) and the so-called Harmony model. The latter model has been a recently proposed method to solve value issues in China, applied in the field of water resources management [3]. We will discuss and compare the two methods in the case of the Yancheng project and argue that the Harmony model is the most appropriate for carrying out this project and possibly similar ones in (or under the supervision of) China in a responsible way.

10.2 Responsible Innovation

10.2.1 Background of the research

Responsible Innovation is a new development concept, partly originating from the field of Sustainable Development. It is also a combination of technological innovation and social responsibility. By now, extensive theoretical and practical research has been done in this domain in several countries. At present, scholars interpret and define the concept of RI from different perspectives. In 2012, the Research and Innovation Council of the European Commission brought out a report entitled 'Ethical and regulatory challenges to science and research policy at the global level' in which it suggests that the concept of RI should be considered in terms of a combination of moral acceptability, risk management and human interests [4]. Schomberg, affiliated with the European Commission, describes RI as a process in which social actors and innovators respond to each other in a transparent, interactive way, taking into account the (ethical) acceptability of the innovation process and its products, sustainability and social approval, embedding the advancement of science and technology appropriately in our social life [5]. The British scholar Richard Owen believes that RI refers to the collective management of science and innovation for the future [6]. The book *Options for Strengthening Responsible Research and Innovation: Report of the Expert Group on the State of Art in Europe on Responsible Research and Innovation* published in 2013 by the European Commission, defines RI as a strategy for research and innovation [7].

Scholarly attention to RI in port development projects has been limited. An occasional paper has been devoted to Responsible Port Innovation [8]. The 16th International Conference on Economics of Infrastructures on 24 May 2013 at the Delft University of Technology had 'Social Responsible Innovation in Infrastructures' as its theme, but port projects were not discussed. However, infrastructural works and port development projects, in particular, are interesting and relevant from the viewpoint of RI, not only because they always involve stakeholders and increasingly occasion public debate and protests, but also because they usually have a double goal, inherent in their public character: the entrepreneurial (or micro-economic) aim and the social (or macro-economic) aim [9]. The entrepreneurial goal is to make the port attractive to users, and to provide a competitive supply of services for its customers; the social goal is to raise citizens' welfare in terms of income, employment, environment, health and other aspects.

10.2.2 Approach of RI

Ravesteijn et al. proposed a five-step procedure for carrying out RI and applied it to the Nanshan Railway project [2]. Their approach consists of the following steps: (1) identify stakeholders; (2) assess impacts of the project to stakeholders; (3) identify diverse values of stakeholders; (4) analyse value tensions among stakeholders; and (5) establish a mechanism to resolve value tensions. In this five-step approach, the core is to deal with value tensions. There are several methods for this, like CBA, Direct trade-offs, Maxi-min, Satisficing, Re-specification and Innovation. The first three methods each suppose a specific form of value commensurability, because of which, they are called optimizing methods [2]. The other three methods are so-called non-optimizing methods. In general, each method has its pros and cons, as shown in Table 10.1.

Cost–benefit analysis is the most commonly used and it is sometimes used in combination with other methods. While these methods focus on the contents, other available methods focus on the process, especially in some form of participation of stakeholders. In the case of the Rotterdam Port, for example, there is a Quality Control Round Table consisting of representatives of all parties and stakeholders; its chairperson is a former Dutch minister [10, 11]. China has a novel approach here, the so-called Harmony model, which will be discussed below.

10.2.3 Harmony model

According to Lao-tzu (571–471 BCE) and Confucius (551–479 BCE), harmony is an orderly, coordinated and natural development trajectory. Chuang-Tzu (396–286 BCE) first proposed the idea about the unity of nature and humans in a harmonious state. In recent years, harmony research is being done to solve the conflicts between a diversity of stakeholders in river basins. The Harmony model refers to human and water systems being in a coordinated state and a virtuous cycle, i.e. improving the water system under the premise of self-maintenance and development, and making water resources available for human survival as well as sustainable social and economic development. The Harmony model is designed not only to systematize data and facilitate the evaluation of the human–water relationship but also to analyse the weaknesses and strengths of water

Table 10.1: Overview of methods for dealing with value tensions.

Method	Advantage	Disadvantage
Cost–benefit analysis	Values are made commensurable on a common scale	Various fundamental and methodological problems with expressing values in money
Direct trade-offs	'Best' option chosen	Making trade-offs may not always be (morally) acceptable
Maxi-min	Selects a design that scores best on the worst criterion	May sometimes lead to irrational choices
Satisficing	The selected alternatives meet thresholds	How to set thresholds in a rational and acceptable way?
	No direct trade-offs between the criteria	May sometimes lead to irrational choices
Judgement	Might solve value tension by judgement and (re-)specification	Re-specification may lead to an unacceptable weakening of moral obligations
Innovation	Can lead to alternatives that are clearly better than all of the present alternatives	Does not solve the choice problem in many cases

resources management and to guide decision-makers to create targeted planning [3].

The Harmony model is a complete process that includes identification, prevention, analysis, processing, and learning and feedback regarding conflicting values. The model includes four subsystems: inspection, prevention, solution and service. The inspection system is responsible for tracking the implementation of the project to identify problems and inform the stakeholders. The prevention system is responsible for controlling the various incentives, environments, orientations, objects and values in the process of the implementation of the project; it aims to nip value conflicts in the bud. Although project managers can take some measures to reduce value tensions, such conflicts are inevitable throughout the life cycle of the project and they

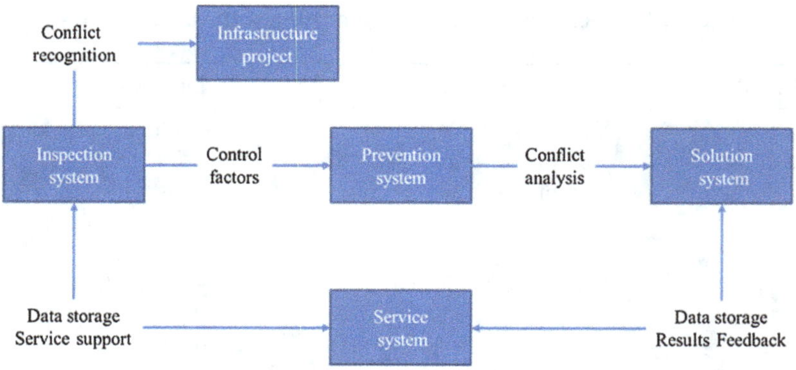

Figure 10.1: Flowchart of Harmony model.

should be dealt with in an appropriate way; otherwise, it will affect the smooth progress of the project. The solution system is mainly used for the analysis and resolution of remaining and other conflicts. The service system is mainly responsible for the management of the database and to provide technical support, while it feeds back the conflict regulation results to the stakeholders. A flowchart of the Harmony model can be found in Figure 10.1. The Harmony model can also be described as a step-by-step plan: (1) identify stakeholders and their values; (2) analyse the value tensions; (3) establish mediation and supervision committees; (4) resolve the value tensions; and (5) store experience.

10.3 Yancheng Port extension

10.3.1 Introduction

With the rapid development of China's economy, an increasing number of ports have been constructed and expanded [12]. Among them, Yancheng Port has been very important. Yancheng Port is a national first-class port as approved by the State Council in 2007 and its extension is one of the five cross-century transportation strategic projects in Jiangsu province [12]. Yancheng Port is located in the middle coastal areas of Jiangsu, covering an area of 500 square kilometres. The geographical advantages of Yancheng Port are evident since it is 250 nautical miles to Shanghai, 420 nautical miles to Bushan and 350 nautical miles away from the Japanese port of Nagasaki. Meanwhile, Yancheng Port is closely linked with the Coastal Highway, Xuda Highway, Ningjingyan Highway, Xinchang Railway, the Beijing–Hangzhou Grand Canal and Yancheng Airport, forming a three-dimensional large traffic pattern.

The Yancheng Port marine economic development zone is located in Yancheng Port – a construction area of 90 square kilometres. The whole development zone mainly consists of Beigang (North Port), Zhonggang (Middle Port), Nangang (South Port) and Harbor New City. The first phase of the project (Beigang and Zhonggang) was put into operation in 2010. Shanghai International Port Group (SIPG) Co. Ltd. won the contract for the Yancheng Port Phase II Project in 2015, which mainly provides for the construction of the Nangang area. The total cost of this project was around US$175 million with a construction period of 16 months from September 2016 to January 2018, covering a land area of about 13 square kilometres. The project includes two 20,000-tonne cargo terminals, a 50,000-tonne cargo berth and a Marine Chemical Industry Park. Expectations are that the entire port throughput will reach 30 million TEU in 2020. Figure 10.2 shows a planning map of this project. We will further analyse the project with the previously mentioned five-step approach, revealing and addressing some of the value tensions.

10.3.2 Stakeholders

There are multiple actors involved in this project, including government agencies, companies and non-governmental organizations (NGOs), all of them referred to here as 'stakeholders'. The main stakeholders are the Provincial Government of Jiangsu (PGJ), the Municipal Government of Yancheng (MGY), Shanghai International Port Group (SIPG) Co. Ltd, marine industries, local residents and NGOs. These stakeholders have different power positions,

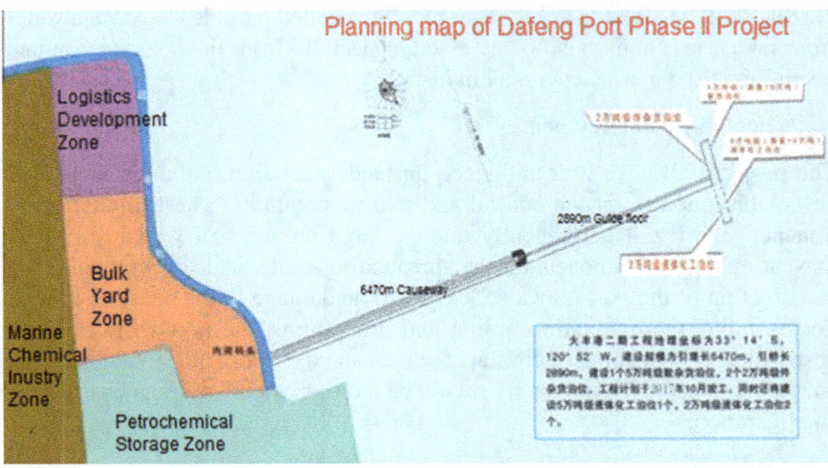

Figure 10.2: Planning map of the Yancheng Port extension project.

while they also differ in other respects, especially their value orientations and the degree to which they profit or suffer from the project, as specified below.

10.3.3 Impact

10.3.3.1 Environmental impact assessment

An environmental impact assessment (EIA) is a formal process consisting of gathering data and evidence, assessing environmental effects, developing mitigation measures and making recommendations to refine plans or programmes in view of the predicted environmental effects, and monitoring the significant effects of implementing the plan or programme [13]. Through it, sufficient awareness of environmental problems will be provided to decision-makers. Stakeholders like local residents and NGOs can make a stand for the environment, whether they directly suffer from negative effects or not. It has been legally compulsory to perform an EIA for big projects in China since 2002 in order to achieve Sustainable Development [13]. The Yancheng Port project mainly consists of two sub-projects, namely Port Extension and the Marine Chemical Industry Park. The project will have significant impacts on the environment as well as some environmental risks that will be discussed further. The impacts on the environment would exist during the construction and operation stages. The key environmental impacts of Yancheng Port are as follows.

- Water environment

Yancheng Port is a deep-water port. Due to the special geographical structure, the tides will be affected after the project. Suspended particles in sewage water from industries should also be assessed, especially from the construction and operation of the marine chemical industry.

- Ecological environment

The project will have little influence on land vegetation, although there will be notable impacts on the coastal and marine creatures. The exploitation of Yancheng Port will permanently remove large amounts of wetlands, which may bring negative impacts on the surrounding ecological environment. The construction of the port may also lead to some damage to the coastal ecology. For example, the excavation, piling and dredging of the port will definitely destroy the living habitat of coastal creatures and fish roe. Thus, living habitats, animal numbers and biodiversity should be monitored both during construction and operation.

- Pollution

Pollution, including air pollution, noise pollution and solid waste, is another main factor that could jeopardize the environment. Air pollution mainly comes from the dust particles released during construction and from the exhausts of machinery, loading and unloading of cargo and industries during operation. Besides, the greenhouse gas emissions cannot be ignored due to the huge energy use. Noise pollution mainly comes from traffic and machinery during construction and cargo handling during operation. Solid waste generated by the port is supposed to be disposed of by the corresponding department of Yancheng City. In this way, solid waste will have little negative impact on the environment of this region.

10.3.3.2 Social impact assessment

The construction and operation of Yancheng Port, as a project lasting for years, would inevitably have considerable influence on society, including the economy and the well-being of local residents. Opportunities could be created, but they would also cause harmful impacts. For instance, the project might bring some changes in the local population structure and industrial diversification. In addition, there is a trend that affected individuals or groups like local residents would play an increasingly significant role in the decision-making of the project. Consequently, identifying and managing social issues are essential for the government of Yancheng City and Shanghai International Port Group. A social impact assessment (SIA), in this case, would help to understand the social consequences and hence contribute to the mitigation of possible negative effects and improve the overall outcomes for local communities and developers.

- Public health and well-being

As mentioned in Section 10.3.3.1, the project could have negative impacts on air quality. People who stay near the development zone of Yancheng Port would be exposed to harmful emissions during the construction and operation stages. The plan of the Marine Chemical Industry Zone would then further endanger local residents' health and well-being due to industrial emissions.

- Demographic change

The construction of Yancheng Port would require a considerable number of workers, which the local workforce simply could not meet. An influx of temporary workers from other cities could be foreseen in the process of constructing the Nangang area. Some may become permanent workers for the industries located in the area and for later port activities.

- Economic change

A substantial amount of investment would be unavoidable for such a large infrastructure project. While it is supporting the port development, it would tend to increase the local price level. More expensive goods might be a burden for local residents as their salaries may not increase proportionally at the same time.

- Resettlement

The construction of Yancheng Port would inevitably involve the resettlement of local residents. It would not be an easy task to reallocate them to other places since they might refuse to move. In addition, the security and living conditions of new locations have to be thoroughly examined.

- Infrastructure development

More frequent economic activities and an influx of workers would stimulate further improvement of infrastructure, such as transportation or other public service sectors. It would consequently lead to an upgrade of the residents' living standards.

10.3.3.3 Risk assessment

Risks are uncertain events or conditions that, if they occur, could have a positive or negative effect on the project objectives. Considering the designed Marine Chemical Industry Zone and the high population intensity, a risk assessment estimating the potential consequences is essential. The assessment will involve not only risk assessment experts but also construction companies, project management teams, technical staff, specialists, local citizen communities and other relevant stakeholders. As EIA and SIA can provide inputs for consequence analysis and scenario development, the risk assessment experts will have to keep in close contact with EIA and SIA experts.

Risks regarding natural disasters should be surveyed. Yancheng is located on the west coast of Huanghai Bay, which is part of the Pacific Ocean. Yancheng has a humid subtropical climate with four distinctive seasons. The rainfall often shows a strong summer peak and typhoons occur frequently. The impact of typhoons depends on their magnitude with a catastrophic negative impact as the worst scenario. Yancheng is not located in any earthquake belts. Though earthquakes may occur, their magnitude is predicted to be too small to make any serious impact on the port.

Risks concerning the construction process of the port are as follows: breaking down of funds; the need for special equipment; land disputes; local communities raising objections; unexpected budget overruns in key materials; functional and technical labour units not available; and damage to historic/cultural sites and endangered species. Among them, land disputes and local community protests are serious risks. The consequences of them are far-reaching, including delays of the schedule and time spent waiting. Risks regarding spending cannot be neglected either. During construction, special equipment and key materials should receive cautious attention.

Risks during the operation are also important. These risks are fire, explosion danger, oil/chemical leakage, change of market conditions and change of environmental restrictions. Fire and explosions might cause significant accidents. The explosion accident in Tianjin Port indicates that safety management of human-induced factors is necessary. From an environmental viewpoint, pollution is unavoidable. Among others, oil or chemical leakages occur occasionally and can cause critical impacts on both water and land ecosystems. Moreover, with the government's continuous efforts for environmental issues and increasing public awareness, environmental restrictions are becoming increasingly severe. Thus, these risks should be considered and flexibility should be kept during the design process. The change of market conditions is important as well. A recession in the world economy, competition from other ports or bankruptcy of companies may lead to a situation in which the port cannot achieve its designed goals and benefits. Consequently, a good selection of contracted companies and estimates are essential.

10.3.4 Values

10.3.4.1 Values and tensions

All stakeholders have their own goals and interests in relation to the Yancheng Port project. Table 10.2 shows the results of a value analysis for each stakeholder.

In order to find value tensions, one must first look at the stakeholder's values and the possible impacts that might affect those values. Thereby, the following two value tensions were selected for consideration.

Table 10.2: Values of stakeholders.

Stakeholders	Economic	Environmental	Social
PGJ	Economic growth		Social stability
MGY	Economic growth		Social stability
SIPG	Profitability		
Marine industries	Profitability	Environmental friendliness	
Local residents		Environmental friendliness	Liveability Health
NGOs		Environmental friendliness	

- The tension between social development and economic growth in the case of local residents who need to be resettled.

When resettlement takes place, people have to start all over again, from finding a new economic activity to sustain their families, to changing their customs according to their new life. These changes in their lives obstruct their social development, which is the price to be paid for higher economic growth.

- The tension between profitability and environmental friendliness, especially when it comes to air and noise pollution.

Worsening air quality will affect people's way of living. When measures have to be taken to reduce air and noise pollution, part of the profits must be given up for the required extra investments.

10.3.4.2 Cost–benefit analysis

In order to abridge value tensions, several methods have been discussed, including Direct trade-offs, Maxi-min, Satisficing, Judgement, Innovation and Cost–benefit analysis, which is most commonly used. CBA can be conducted to provide an overall evaluation of the Yancheng Port project to determine whether it would be profitable or not, which might help to solve the value conflicts between stakeholders [14]. To ensure the CBA analysis is as accurate and comprehensive as possible, it should take a broad view of the costs and benefits, including direct and indirect factors, monetary and non-monetary items and short-term and long-term effects.

Analysis from different perspectives reveals that the Yancheng Port project involves plenty of costs and benefits. Direct costs include the cost of relocation fees, construction, operation, maintenance, tax and loss of wetlands. Indirect costs include loss in existing marine industries and environmental impacts from construction, such as air quality and noise. As for the benefits of Yancheng Port, direct benefits include the gains from the operation of the main business of the port and improvement in transportation and logistics. Indirect benefits embrace social impacts, especially improvements to resident incomes, employment, tourism, local infrastructure and the hinterland economy.

Cost–benefit analysis and the other methods focus on the content. Process methods have also been put forward, and the Chinese Harmony model is a case in point.

10.3.4.3 Harmony model

In the Yancheng Port project, the Harmony model gives rise to the following actions: building a conflict resolution system; identifying value tensions; preventing conflicts; analysing and handling conflicts; and feedback and

learning. These actions have to be taken in the framework of the earlier indicated step-by-step plan, creating the following procedure:
Step 1. Identify the Stakeholders and their Values. As mentioned previously, the main stakeholders in the Yancheng case are PGJ, MGY, SIPG Co. Ltd, marine industries, local residents and NGOs. Table 10.2 shows their values.
Step 2. Analyse the Value Tensions. Section 10.3.4.1 gives two main value tensions: the tension between social development and economic growth in the case of local residents who need to be resettled, and the tension between profitability and environmental friendliness, especially where it comes to air and noise pollution. These two tensions would be the focus of the inspection system in the Harmony model.
Step 3. Establish Mediation and Supervision Committees. Considering the necessity of solving the potential and unpredictable value tensions, the Harmony model provides the establishment of committees responsible for managing or supervising the value tension resolutions involved during the Yancheng Port project. The project managers (PGJ and MGY) should be in the first coordination meeting to install the value tension resolution system. The value tension solution system consists of three parts:

1. Value Tension Processing. The scope and size of the conflict handling system vary with the progression of the project and related responsibilities. The number of conflicts diminishes at higher processing levels, although simultaneously the intensity of the conflicts increases.
2. Setting up the Dispute Adjudication Board (DAB). The value tension between the construction unit (SIPG) and the local residents seems the main conflict among the stakeholders. Therefore, both the construction unit and the local residents jointly appoint the members of the committee as impartial third-party mediators.
3. Setting up an Emergency Team. Construction units, local residents and government agencies are represented in this team, which has to deal with important emergency conflicts. The representatives of government agencies constitute the leadership in these teams.

Step 4. Resolve the Value Tensions. In the Harmony model, value tensions would be inspected, prevented, solved and assisted in different systems. The four systems – inspection system, prevention system, solution system and service system – and the stakeholders involved in each system are shown and analysed in Table 10.3. It is different from Table 10.2 as the stakeholders in Table 10.3 include DAB and the emergency team. The tasks these systems are performing can be detailed as follows:

Table 10.3: Systems and stakeholders in the Yancheng Port extension project.

Systems	Stakeholder involved	Concrete practice in Yancheng Port
Inspection system	PGJ, MGY, SIPG	Monitor the whole process of Yancheng Port projects; trace the status of the conflict in the system
	DAB	Predict and report on the value conflicts; gather full information about a conflict; number and mark each value conflict
	Marine industry, local residents, NGOs	Submit their value issues through the network terminal; trace the status of the conflict in the system
Prevention system	PGJ, MGY, SIPG, DAB	Control various conflict factors; trace the sources of perceived conflicts; take measures
Solution system	DAB, the emergency team	Determine the urgency level; collect information from different stakeholders; promote the achievement of consensus
	Conflicting parties	Submit acceptable solutions; compromise and implement the solution
Service system	Conflicting parties	Evaluate each other's cooperative attitude and conflict handling ability; give feedback and learn by doing
	DAB, the emergency team	Prevent the recurrence of conflicts; achieve a harmonious state for the conflicting parties and the whole Yancheng project

1. Inspection system. The identification of value tensions includes two parts. On the one hand, project managers monitor the whole process of port projects. When a possible value conflict is predicted, a report is submitted from the network terminal login server and a number similar to the order number is obtained. Conflict situation reports include descriptions of existing conflicts, possible conflicts and conflicts involving stakeholders. After the conflict is processed, project managers can use this number to mark the results of the conflict resolution. On the other hand, the project participants can gather full information about a conflict, its description and stakeholders involved and are able to determine whether the conflict is urgent. They can submit their issues through the network terminal and get a similar order number. The stakeholders can use this number to log on to the server to query the

status of the conflict. On the website of the Yancheng Port Authority, there is a public involvement page, including director mailbox, internet survey, citizen opinion collection and investigation report. It is a good gesture to listen to the resident side. But most concerns there are related to economics. If there can be more value concerns, the inspection of value conflicts would be easier.

2. Prevention system. In the process of construction, project managers can reduce the occurrence of conflicts by controlling various factors that may lead to value conflicts inside and outside the project and carry out an analysis of perceived conflict situations to find their sources. After the central server sends the conflict situation report to the stakeholders, the project manager determines the measures to be taken according to the characteristics and causes of the conflict. He can also avoid the conflict, if it is not serious enough, or steer developments in a positive direction. In Yancheng Port, one potential value tension is between profitability and environmental friendliness. In the prevention system, PGJ, MGY, SIPG and DAB could identify the concerns about environmental friendliness, and they should actively publish related information to the NGOs and residents. The publication of the 'Marine Environmental Impact Statement on Engineering Structure Upgrade and Reform of Bulk Cargo Terminal Extension II in Dafeng Yancheng Port' is a good example. Some surveys, such as questionnaires that Yancheng Port has already designed and posted to the residents, are also very helpful in promoting communication between stakeholders and preventing value conflicts.
3. Solution system. When value tensions occur, it is first necessary to determine the urgency of the conflict and, in the event of a serious and urgent situation, the server will inform the emergency team directly to handle the conflict. In the event of a general conflict, the server forwards conflict reports sent by the project participants to the stakeholders, while the stakeholders submit the solutions they can accept to the server. And then the project manager makes a summary to determine whether the parties can achieve an agreement. If there is such an agreement, the manager will inform the conflicting parties about the solution. If the views are inconsistent, the manager will involve higher management levels. If consensus is reached, the conflict parties are notified to implement the solution. If not, the problem will be transferred to the next level. If necessary, the problem will be dealt with by third-party mediators.
4. Service system. After the resolution of a value issue, its causes are analysed to prevent the recurrence of such conflicts, while the conflicting parties can evaluate each other's cooperative attitude and

conflict handling ability. This kind of evaluation is a kind of feedback and learning on the basis of the result of the conflict. It can happen that the conflict will not be exposed and dealt with in time, so it is also important to monitor the conflict and achieve a harmonious state.

Step 5. Store Experience. The processes and resolutions of each solved value tension should be summarized and stored in the Harmony model database, which establishes an inventory of examples for future possible value tensions. Once similar value conflicts occur, previous tensions could provide a reference. If an unpredictable and unprecedented value tension turns up, the resolution experience should be added to the inventory. With the accumulation of experience and cases, the Harmony model could be progressively more efficient.

10.3.5 Discussion

The CBA method mainly tackles value oppositions from the economic point of view. If the benefit is higher than the cost, the solution will be preferred, if not, the project will be postponed or even no longer considered. All relevant considerations in a CBA can be expressed in a common currency, such as the euro or RMB [8]. Because all values are measured on a common scale, the CBA assumes ratio measurability and value comparability. The advantage of this assumption is that the best alternative can be selected by the option score in a common measure: the monetary value representation range. If we want to apply the CBA for value tensions, we somehow need to express values, such as security and sustainability, in terms of gain and loss. Take security as an example: it is difficult to value the varying degrees of security that various designs provide in money and to do so in a reliable and uncontroversial manner. Therefore, based on the above considerations, CBA may not be able to solve the value oppositions in a project alone, and it may even not be the best way for the Chinese context.

Compared with other methods of dealing with the value tension, the Harmony model is a process method. The entire project is considered as a holistic system that dynamically monitors the value tensions between stakeholders within the system, prevents possible conflicts and provides feedback on the conflicts that have occurred to avoid such conflicts in the future. The model makes government agencies the main coordinating institutions, but at the same time, it establishes a third-party mediator, so as to ensure the quick and effective settlement of the vast majority of conflicts, which will greatly improve the efficiency of port projects. The model has important guiding significance to address the value problems that China port constructions may face.

10.4 Conclusion

The core objective of a port project is to meet the needs of the various stakeholders while completing the project. The success of such projects is not only the responsibility of project managers and builders but also of all stakeholders involved. The complexity of port projects determines that the realization of their objectives requires the participation of a wide range of stakeholders or interest groups whose needs and expectations vary from project to project. Therefore, in the course of a project, conflicts inevitably emerge during design and implementation and conflicts may occur throughout the full life cycle of a port project. Therefore, it is necessary to study and resolve the value tensions of the project developers and all other stakeholders.

Responsible Innovation has been put forward as a new development concept suitable for projects that entail value conflicts, and a practical step-by-step approach has been suggested, consisting, in particular, of Stakeholder Analysis, Impact Assessment and value conflict resolution. Consequently, this chapter has discussed the so-called Harmony model and applied it to the presently planned Yancheng Port Extension Project.

The following research results have been obtained. First, we have discussed the Harmony model as a process method for resolving value conflicts, alongside CBA as an example of a method focused on the contents. We argued that content methods cannot be the only solution and we showed that the Harmony model is suitable for solving value tensions in China. We think such process methods can be especially worthwhile in the situation of a developing society and economy, not least because they can be based on a country's cultural heritage. For China, the Harmony model, with a history going back to ancient times, is a good example, which deserves to be followed, although adapted to a country's special conditions.

Second, based on the analysis of impacts and stakeholder interests and values, this chapter specified characteristics, types and approaches of value conflicts in the Yancheng case and, consequently, it constructed a harmonious model of stakeholder involvement for the project. Third, we showed that project managers can minimize the occurrence of conflict through analysis and resolution, aimed at controlling the environment, incentives and other factors in dynamic processes. When a conflict occurs, the manager can take different approaches depending on the importance of the conflict at different stages of the project's life cycle.

In the proposed Harmony model-based Yancheng Port development, its Responsible Innovation mode is mainly guided by the government's macro-control and coordination, although with the participation and consultation of different stakeholders in order to achieve a collective responsibility for the entire port construction process. In view of China's present economic and social development, governmental intervention based on the Harmony model would be conducive to achieving overall effectiveness. The model also has a great guiding significance for port construction and expansion across borders in the One Belt, One Road project.

References

[1] Development and Reform Commission, 2015 Annual Report of Development and Reform Commission of the People's Republic of China. Development and Reform Commission, Beijing, 2016. (In Chinese.)

[2] Ravesteijn, W., He, J. & Chen, C., Responsible innovation and stakeholder management in infrastructures: The Nansha Port Railway Project. *Ocean & Coastal Management*, 100, pp. 1–9, 2014.

[3] Zhang, J., Tang, D., Ahmad, I. & Wang, M., River–human harmony model to evaluate the relationship between humans and water in river basin. *Current Science*, 109(6), pp. 1130–1139, 2015.

[4] Ozolina, Z., Mitcham, C., Schroeder, D., Mordini, E., McCarthy, P. & Crowley, J., Ethical and regulatory challenges to science and research policy at the global level. Directorate-General for Research and Innovation, European Commission, Brussels, 2012.

[5] Von Schomberg, R., A vision of responsible research and innovation. In: Owen, R., Bessant, J. & Heintz, M. (eds), *Responsible Innovation: Managing the Responsible Emergence of Science and Innovation in Society*. Wiley, Chichester, pp. 51–74, 2013.

[6] Stilgoe, J., Owen, R., & Macnaghten, P., Developing a framework for responsible innovation. *Research Policy*, 42(9), pp. 1568–1580, 2013.

[7] Van den Hoven, J., Jacob, K., Nielsen, L., Roure, F., Rudze, L. & Stilgoe, J., Options for Strengthening Responsible Research and Innovation, European Commission, Brussels, 2013.

[8] Van den Hoven, J., Wildschut, E. & Bakker, E., Responsible innovation at port cities. *Bollettino del Dipartimento di Conservazione dei Beni Architettonici ed Ambientali* 12(1), pp. 765–770, 2012.

[9] Grigalunas, T.A., Luo, M., Chang, Y., Comprehensive Framework for Sustainable Container Port Development for the United States East Coast, Year One Final Report, URI-TC Project, No. Grant 536106, 2001.

[10] Dix, J., Better Scope Management during the Initiative, Design and Construction Phase of Infrastructure Projects, Master's Thesis, Delft University of Technology, 2010.

[11] Lichtenberg, J.J.N., Maak innovatie hsl en Betuwelijn transparant. *Cobouw*, 25 January 2005.

[12] Wang, C. & Ducruet, C., New port development and global city making: Emergence of the Shanghai-Yangshan multilayered gateway hub. *Journal of Transport Geography*, 25, pp. 58–69, 2012.

[13] Dalal-Clayton, D.B. & Sadler, B., *Strategic Environmental Assessment: A Rapidly Evolving Approach*, International Institute for Environment and Development, London, 1999.

[14] Orth, K., Robinson, R., & Hansen, W., Making More Informed Decisions in Your Watershed When Dollars Aren't Enough (No. IWR-98-R-1), Army Engineer Institute for Water Resources, Fort Belvoir, VA, 1998.

CHAPTER 11

Onshore power versus fuel switching in the mouth of the Pearl River with reference to the Port of Nansha

Aida Cahyaningwidi, Jannis Langer, Praveen Sridharan & Frank Vollering

Faculty of Technology, Policy and Management, Delft University of Technology, The Netherlands

Abstract

By analysing a set of innovative technologies and suggesting policies on how to implement them accordingly, we show how the Port of Nansha, located in the mega-city around the mouth of the Pearl River, can establish Sustainable Development by means of Responsible Innovation and become a paradigm of port sustainability. Two main questions will be dealt with: how can sustainable and responsible development of the Port of Nansha be advanced through innovative energy technologies and which policies and institutions could support the implementation of such technologies in view of local stakeholders and their values?

Keywords: The Port of Nansha, responsible development, sustainable energy technologies

11.1 Introduction

President Xi Jinping's brainchild and China's bold undertaking of rejuvenating the Silk Road under the Belt and Road Initiative (BRI) could redefine the geopolitical landscape. Through this programme, China aims to interconnect transport corridors and port facilities. This will go a long way in boosting trade, improving security and also cementing China's place as a political stronghold in the world while providing sharp economic competition against global leaders like the United States [1].

Asia's importance in the international economy continues to rise as it accounts for one-fourth of world trade. However, its potential has not been realized due to considerable bottlenecks in infrastructure, transport and logistics. Based on its own interests, China has taken the responsibility to seize the centre stage in Asian affairs. Therefore, the BRI is a promising but challenging task that aims to boost China's western and central provinces through the inflow of new jobs and business opportunities. Another effect is the forging of new diplomatic ties, as China provides a helping hand to other countries involved in the BRI, namely by funding and policy support [2]. Although the BRI comprises many benefits, it cannot be excluded that these emerge at a cost. Recently, increased attention has been paid to Responsible Innovation, which requires the consideration of all stakeholders either involved in or affected by decision-making, thus curbing the potential of conflicts. That this mindset is not far-fetched is shown by incidents that occurred in Shatian in 2013. The loss of employment of local villagers due to semi-legal and illegal corporate activities induced riots, which entailed acts of violence and disorder [3]. This should be motivation enough to include Responsible Innovation as an overarching motive in BRI projects, one of which encompasses the expansion of the Port of Nansha.

The Port of Nansha is one of the four constituents of the conglomerate of the Port of Guangzhou, which stretches over a 400 km coastline. Compared with the other three members, namely Inner Port, Huangpu Port and Xinsha Port, the Port of Nansha is the most prominent and productive one, with more than 60% of activities of the Port of Guangzhou conducted there [4, 5]. Being part of the Greater Pearl River Delta (PRD), the Port of Nansha arose as one of the largest container ports in the world with 9.52 million TEUs in 2012 and resembles a paramount juncture, not only of the Chinese but also the global economy. China recognizes the significance of the Port of Nansha, and its development lingers at the top of their priorities, as, for example, seen by the Nansha Port Railway Project, the implementation of which started in 2013 [3].

Although the Port of Guangzhou's and consequently the Port of Nansha's importance for both foreign and domestic trade is undisputed [6], all these activities and endeavours come at a price. The Port of Guangzhou stands out in terms of ship emissions and is ranked first throughout China in all types of emissions in 2013, from SO_x to NO_x and CO_2, particulate matters (PM) and volatile organic compounds (VOCs) [7]. In contrast to the Port of Shenzhen, another major port of China, the Port of Guangzhou is reluctant to issue action plans towards emission mitigation. At least, policies were recently introduced both on the national and provincial level to counterbalance the consequences of port activities. For instance, in 2015, the Chinese Ministry of Transport established stringently regulated emission control areas to reduce sulphur emission from shipping [8]. The ministry also issued the National Five-Year Plan for Environmental Protection and the National Five-Year Plan for Energy Saving and Emission Reduction [9]. On the provincial level, the government of the Guangdong Province introduced the Guangdong Green Action Plan (2014–2020) [7, 9] as well as the Provincial Air Pollution Prevention Action Plan [9]. However, despite recent institutional activities on national and provincial levels, the Port of Guangzhou itself still lacks specific regulations.

In regard to the BRI and the progression of the Port of Nansha, this might pose severe obstructions both on national and global levels. On the national level, the potential for resistance comes from local communities, as decision-makers rarely include them in a suitable way during project implementation. In this respect, China signed multiple intergovernmental agreements, joint statements and many other documents with countries like Thailand, Pakistan and South Africa [10]. Thus, China has a responsibility to leap forward as a role model, not only to maintain and consolidate the trust of its partner countries but also to ease the very obstruction of involved, local stakeholders. With all these insights, a port as integral to domestic and global trade as the Port of Nansha could become a paradigm of sustainable and responsible port development and serve as an example for other ports worldwide, independently of their involvement in the BRI. For this to happen, both technological and institutional change is required, not only on the national and international levels but also on the local level of the Port of Nansha itself.

By analysing a set of innovative technologies and suggesting policies on how to implement them accordingly, in this chapter, we will show how the Port of Nansha can reduce emissions and establish Sustainable Development by means of Responsible Innovation and so become a paradigm of port sustainability. Two main questions will be dealt with here, which not only focus on technical solutions but also the required economic, environmental and societal conditions, related, for example, to the stakeholders and their values: how can sustainable and responsible development of the Port of Nansha be advanced through innovative energy technologies and which policies and institutions could support the implementation of such technologies in view of local stakeholders and their values?

First, we will explain our approach, followed by an overview and assessment of the available energy technologies and their economic and socio-environmental characteristics. Second, we will focus on the stakeholders and their values. Third, we investigate and discuss what can be done to solve value conflicts and implement technologies in ways acceptable and desirable to the stakeholders. Finally, we present our conclusions and show ways to proceed with this line of research.

11.2 Structure and approach

11.2.1 Specific questions and methods

A patchwork of multiple methods and frameworks has been used to answer the research questions, focused on the following sub-questions:

1. What energy technologies are available to contribute to a more sustainable and responsible operation of the Port of Nansha and what are their economic and socio-environmental characteristics?
2. How can a portfolio of available energy technologies be composed on the basis of predefined criteria and what are the timescales for the individual technologies to be deployed in the most feasible fashion?
3. Who are the local stakeholders affected by the economic, environmental and social implications of technology implementation and what are the possible value conflicts among these stakeholders?
4. Which policies and institutions could reconcile stakeholder value conflicts to improve the feasibility of the technology portfolio implementation?

Sub-question 1 serves as an introduction and rests on the findings of a general study of current literature and provides a global understanding of the technical, economic and socio-environmental aspects of the technologies concerned (Sections 11.3.1–11.3.3). The second sub-question synthesizes these understandings in a comparative manner. Ultimately, this results in a decision matrix for finding the best combination of technologies to ensure sustainable and responsible development of the Port of Nansha (Section 11.3.4).

First, multidisciplinary domains are defined on the basis of which technologies are compared. These domains are provided with weighting factors, which represent the desirability and preference of specific criteria. Then the technologies are rated according to the defined aspects on a predetermined scale, depending on how well a technology meets the respective criteria. Points are multiplied with the assigned weighting factors and consequently aggregated for each technology. The technologies with the highest total number of points are considered the most suitable options [11].

The criteria for the assessment of the available technologies have been retrieved from the paper by Van Alphen et al. [12]. Their approach was chosen as it covers different, interdisciplinary domains and proved its utility for renewable energies in developing countries like the Maldives. It comprises:

1. Economic Profitability: capital and operational costs, return of expenses, cost reduction potential, etc.
2. Sustainability: lifetime of technologies, emission mitigation, impact on society and environment, etc.
3. Renewable Energy Supply: proportion of renewable energy used/usable by the technology
4. Compatibility: consistency with existing technologies, extent and intensity of changes required to implement the technology

5. Simplicity: the easiness of studying, constructing, operating and maintaining the technology

The criteria are rated from 0 (not applicable at all) to 5 (very applicable) in intervals of 0.5. Moreover, the weighting factors (WF), with which the ratings are multiplied, go from 1 (less important) to 3 (very important) and are available in intervals of 0.5. The averaged unweighted scores are abbreviated to AUS, while the average weighted scores are called AWS.

We are well aware of both the benefits and shortcomings of this decision-making tool. On the one hand, it enables the quantification of preferences and the articulation of criteria and their significance. By grading and weighting, the composition of a technology portfolio is well rounded and vindicated. But on the other hand, the method is fundamentally compromised by subjective bias: the determination of both the weighting factors and the ranking of the technologies depends on our personal decisions. Knowing this, we try to base our assessments on thorough knowledge based on and aligned with current literature. Additionally, by peer-reviewing each other's decisions, we strived for as much objectivity and integrity as possible.

After a general assessment of the relevant technologies, our focus shifts to the Port of Nansha (Section 11.4). A stakeholder analysis (sub-question 3) will shed light on the economic, environmental and social impacts of each technology on local stakeholders. Here, the methodology of Ravesteijn, He and Chen [3] is used. In their paper, they perform a stakeholder analysis in the same place, namely the Port of Nansha. Five steps are taken: (1) Identification of Stakeholders; (2) Impact of Project(s) on Stakeholders; (3) Identification of Stakeholder Values; (4) Value Conflicts; and (5) the Constitution of Governance Mechanisms for Value Conflict Resolution.

In addition, fragments of the framework of the Multilevel Perspective (MLP) of Geels [13] are used for further structuring our argument. Stakeholders from different domains interact with each other, which can be cooperative or competitive [13]. Here we will address the latter type in the form of value conflicts. For that purpose, we will draw on the available literature and corporate and public websites.

Finally, the last sub-question draws partly on a literature study again. Here, the last of the five steps of Ravesteijn, He and Chen's methodology is performed, the constitution of a governance mechanism to resolve the value conflicts (Section 11.5). Thereby, possible policies and institutions are discussed that might reconcile detected stakeholder value conflicts and thus improve the feasibility of the technology portfolio implementation. In conclusion, the findings are summarized and the main research question will be answered.

11.2.2 Two types of analysis

This contribution bears the character of an advisory report for decision-makers on the Port of Nansha. We provide suggestions for the port authorities to establish Responsible Innovation and Sustainable Development from an energy technology perspective. In Section 11.3, we carry out an expert analysis, in which we assess a series of three sustainable energy technologies on technical, economic and socio-environmental peculiarities and propose a technology portfolio composed of the most suitable set of options. These insights are generally applicable. In this regard, we consider ourselves – the experts – as stakeholders as well. In Section 11.4, we switch to a more detailed scope and adjust our reasoning in accordance with a stakeholder analysis. We search for the stakeholders involved in and affected by the implementation of our set of technologies and identify stakeholder values and value conflicts between them. Here, our insights specifically concern the Port of Nansha and the surrounding district. Here, the experts are not included in the analysis. The policy recommendations in Section 11.5, intended for every stakeholder group to reconcile the implementation of technologies considering their value conflicts, combine the findings of both types of analysis to tackle the problems in the Port of Nansha (see Table 11.1).

11.2.3 Disclaimer on the cultural bias of this report

We are aware that our assessments may be fundamentally biased by our Western perceptions and that basic assumptions like liberal markets do not always apply to Chinese conditions. Our work is strongly inspired by the book *Why Nations Fail* by Acemoglu and Robinson [14], who praise the significance and necessity of inclusive institutions for development while criticizing extractive ones. We firmly believe that our recommendations for political reform and a consequent shift to more inclusive policies, resulting in more rights for less powerful stakeholders and a more liberal society, are useful, though we also realize that our arguments might have to be adjusted to concur more with the underlying perspectives of harmony prevailing in

Table 11.1: Differences between expert and stakeholder analysis.

Expert analysis	Stakeholder analysis
General	Specific
Applicable worldwide	Only applicable for Nansha
From an exogenous point of view	Focus on a local stakeholder perspective
Experts part of stakeholders	Experts not part of stakeholders

China. For example, instead of 'conflict reconciliation', our contribution could have emphasized 'harmony reinforcement', which basically strives for the same objectives, but from a different angle. Thus, we do not think that our findings and insights are flawed, but that they have to be translated into another conceptual language in order to make them more applicable to Chinese circumstances. For this to happen, we propose further research, basically advocating a more intense inclusion of Chinese experts and stakeholders.

11.3 Characteristics of available energy technologies

This section will briefly discuss the different technologies, namely onshore power, fuel switching and fuel cells on ships, from a general perspective. The technologies have been chosen because they were recently developed and have a promising potential for sustainable and responsible port operations in Guangzhou, as described in [15].

11.3.1 General overview of available technologies

11.3.1.1 Onshore power

With onshore power, the power provision of ships docked in the port shifts from auxiliary diesel engines on the ship to external power provided by the local electricity grid. Thus, the level of emissions at the port is reduced significantly, rendering conventional auxiliary systems on ships obsolete [16]. Onshore power has attracted a lot of attention recently, with more than ten ports worldwide participating, as of 2015, and some Asian ports considering the implementation of the technology in the near future [15].

Nevertheless, myriad details regarding the system and its infrastructure have to be taken into account before the technology can be broadly implemented. For example, both old and new ships have to undergo changes in their electricity management systems. While retrofits can be used for the former, new system requirements like energy management systems and smart meters have to be included in the latter. Ensuring a seamless power transfer is, indeed, complex and requires an overhaul of existing onboard power systems [17]. The types of vessel and traffic in ports also have to be considered. There are a set of typical power requirements for different ships that are important in designing the points of connection (layout, number of shore connections, complexity, etc.). Hence, connecting ships to onshore power resembles a mosaic of several small pieces, both at the demand and the supply side, which needs to be in place to adjust the power delivered to the ships.

To achieve a net reduction of emissions, the provided electricity must originate from renewable energy sources instead of conventional ones from fossil and nuclear fuels. If the latter is used, the emissions are merely shifted from one location to another. In this case, onshore power is returned from a net benefit to a zero-sum technology.

11.3.1.2 Fuel cell ships

Another feasible option for a more sustainable operation of ships – both at sea and docked in the port – are fuel cells, converting chemical energy stored in fuels like hydrogen, natural gas, ammonia or diesel directly into electrical energy [18–20]. Due to their versatility, fuel cells find their place in a wide range of applications in maritime transportation. Hence, the technology is suitable not only for passenger ships, ferries and yachts but also for naval operators [20, 21]. Depending on preference and desired utility, fuel cells are conceivable as clean auxiliary systems, but also as standalone propulsion systems [19, 21]. In the past, projects have been carried out that studied the technical feasibility and operational peculiarities of fuel cells on ships, such as METHAPU, FellowSHIP and e4ships [20]. Benefits of fuel cells include clean operation and emissions as well as reduced noise production [19]. Shortcomings comprise limited lifetime, low durability, high costs, problems with fuel storage and fuel selectivity [18]. Additionally, virtually no infrastructure exists for the required fuel, which implies a high probability of fuel supply and distribution disruptions [22].

11.3.1.3 Fuel switching

Shipping companies usually use heavy fuel oil (HFO) for powering their ships, mainly due to its relatively low prices. A side effect of these heavy fuels is that they exhaust a large number of pollutants when combusted, among others SO_X, NO_X, CO, PM and VOCs. These compounds have a serious impact on the environment and may cause health issues to exposed living beings [23]. According to Lack et al. [24], the contribution of commercial shipping to air pollution has been recognized as significant in recent years. Therefore, to shift to a more sustainable operation at the Port of Nansha, fuel switching could be a solution. Thereby, HFOs are only used in open seas, but when entering the port, fuels are exchanged for a lighter distillate of crude oil containing fewer pollutants. Currently, a number of Western countries have already imposed strict regulations on fuel properties, such as sulphur caps on the open sea. According to Mestl et al. [25] (2013), the maximum limit for sulphur concentration in marine fuels on open seas was lowered from 4.50% to 3.50% by the International Maritime Organization (IMO) in 2012. However, a huge portion of countries in the world still do not regulate the quality of fuels deployed in their ports. In addition, the use of low sulphur and low viscosity fuels might exert several harmful effects on diesel engines, as stated by ABS [26] in their advisory report. This also explains Annex VI of MARPOL, which defines regulations for the prevention of air pollution from ships and international regulation that is accepted worldwide to decrease these emissions.

11.3.2 Economics of considered technologies

11.3.2.1 Onshore power

There are many types of costs involved in onshore power. For fixed investments, there are the costs of ship-side and shore-side infrastructure. Roughly, ship-side modifications range between US$300,000 and US$2 million. Moreover, the costs for the transmission of electricity from the local grid to the shore-side terminal are estimated to oscillate between US$1 million and US$3 million, albeit the underlying costs vary depending on the location of the port and the grid [27].

Onshore power can either be introduced by the development of new infrastructure or the modification of existing facilities. However, a new terminal equipped with onshore power currently costs less than re-designing an old one. In some cases where docking is not possible, ancillary equipment such as barges is required to connect the ships to the grid. This will add to the costs of both ship-side and shore-side connection infrastructure. The operational costs involved in onshore power can be divided into energy costs, labour costs and routine maintenance costs [28]. Additionally, unplanned maintenance costs also have to be taken into account, which can be deduced from experiences with previous projects. For small ports that still have to make the jump into the 'busiest ports' section, this is an ideal opportunity to catch up. Developing onshore power will ensure that ports abide by IMO regulations and will bring more ship traffic to ports.

For ships as well, the costs for retrofits are considerably higher compared with new ships, which could be one of the primary reasons why shipping fleets might not be inclined to adapt to onshore power. Another reason that could influence decisions is the price of the electricity. High prices of port electricity compared with bunker fuel prices are likely an impediment to the wide-scale diffusion of onshore power. For instance, MAERSK claims onshore power is not a cost-effective technology for ships that dock for short periods, at least in US waters. Consequently, onshore power is unlikely to make any headway as long as electricity prices exceed the costs of bunker fuels [29].

11.3.2.2 Fuel cell ships

Due to their relatively early stages of development, most types of fuel cells are far from market penetration and still require further research. Although the relative fuel cell stack costs have been reduced by 60% throughout the last decade [20], upfront investment costs make the technology less competitive compared with alternatives. The most advanced types of fuel cells require capital investment costs of over 5000 $/kW [19]. Meanwhile, alternatives at an earlier stage of development involve even greater costs, like solid oxide fuel cells (SOFC) with around 9000 $/kW at low production volumes [20]. Nevertheless, it is not surprising that fuel cells fail to grasp a substantial market share and instead linger in a niche existence [19].

Financial-economic hurdles do not only originate from intrinsic technical, but also from systemic limitations, typical for any type of innovative technology. The gigantic capital costs for the establishment of a fuel infrastructure strongly inhibit the implementation of fuel cells [30]. Besides that, there are also myriad other criteria affecting fuel cells, like stack and inverter costs [19].

Nevertheless, researchers believe that the costs of SOFC, one of the most promising types of fuel cells, will decline considerably in the future, although it is uncertain to what degree. For example, de-Troya et al. [20] estimate the future costs of SOFC to plunge to 750 $/kW, while Van Biert et al. [19] state that future stack production costs between 150 and 1500 $/kW are conceivable. According to the US Department of Energy (DoE), production costs of 1000 $/kW for fuel cells in general are low enough for entrance markets, while a mature market would require costs of 500 $/kW, with SOFC as the technology most likely to fulfil these requirements [21]. For the necessary full market penetration, public support would be inevitable [19]. For instance, the European Union launched the National Innovation Programme (NIP) for hydrogen and fuel cell technology [30], which supports the further development of these technologies and thus improves their competitiveness.

11.3.2.3 Fuel switching

Costs connected to fuel switching imply higher prices for, among others, cleaner fuels, technical modifications of machinery to allow alternative fuels, fuel storage, maintenance costs and logistics. According to ABS [26], most ship engines have been designed to operate primarily on cheap, purely residual HFO, with occasional use of distilled marine diesel oil (MDO). As the distillation of crude oil is a costly process, MDO is approximately 50% more expensive than HFO, which explains why it is not utilized continuously [31].

The total costs of fuel switching are divided into initial costs and periodical costs emerging during operation. The first type of costs is mainly for the refineries that need large-scale capital mobilization due to the required specialized equipment and accessories [32]. Regarding the latter, ship operators bear the consequences and, henceforth, face a net loss in profits [31]. But the implications of more stringent fuel quality standards are not only economic but also geopolitical. The transition to cleaner fuels might incentivize countries to seize a larger global market share by leapfrogging towards more efficient high-tech refineries [32], thus threatening competitors with outdated technology. Fuel refiners can expect a larger demand for distilled fuels after the introduction of pollution abatement policies, and those capable of serving this demand with standardized fuels will enjoy rising revenues. Those with obsolete technology, unable to meet more rigid restrictions, will suffer from severe losses of profits [33].

11.3.3 Socio-environmental peculiarities of the technologies

Some socio-environmental aspects relate to all of the listed technologies. All of them reduce the exhaustion of compounds like NO_X and SO_X, while CO_2 is considerably mitigated by onshore power and fuel cells, but less by fuel switching. Lower concentrations of hazardous pollutants in the air improve people's quality of life, especially because of the positive health effects, for example, fewer people with asthma [34]. Besides less smog, lower levels of pollutants also tackle the problem of acid rain, which causes harm not only for fauna but also flora, like crops and feedstock. Less pollutant emission benefits the maritime ecosystem as well, for example by the reduction of seawater acidity due to CO_2 absorption [35–37]. Generally, these technologies contribute to global warming mitigation, which would be especially important for the Port of Nansha as a possible paradigm of port sustainability. Each of the three technologies is assessed individually in the following sections.

11.3.3.1 Onshore power

Introducing onshore power implies that major constructions will initially have to be built to establish substations, power lines and other related devices. According to Ravesteijn et al. [3], this phase will cause severe disruptions in local environments. First, local air quality will deteriorate due to higher dust concentrations at the site. Second, the process of laying electricity cables will irreversibly destroy vegetation and agricultural plots, directly affecting farmers. Last, the sand and mud excavated by construction activities could pollute the water of the surrounding areas. During operation, there will be electromagnetic radiation from the power devices [3], with severe health risks especially after longer periods of exposure. Additionally, visual pollution of the grid connection will result in declining value from recreational and local real estate perspectives. On the bright side, the use of onshore power will reduce noise pollution as ships turn off their auxiliary engines when they are docked in the port. Onshore power will reduce the concentration of pollutants at the local level, but at higher levels, it is crucial to employ electricity from renewable energies. If not, onshore power could be devalued as a zero-sum technology, merely shifting emissions from one location to another [15].

11.3.3.2 Fuel cell ships

Ammonia and methane could be used as sources, but here we concentrate on the main one: hydrogen. Far-reaching construction and modification measures are required to build a hydrogen supply and distribution infrastructure, including pipelines, pumps, tanks and many other accessories. Like onshore power, this will inflict considerable disruptions in the landscape, for example, by soil excavation or deforestation. Not only the construction of the infrastructure, but the production of the necessary materials and components involves energy and resource-intensive activities, including the mining and processing of metal ores and various precious metals as well as processes to forge highly specialized materials capable of meeting the demands of hydrogen transportation [38]. Even with good materials, hydrogen leakage would be a serious socio-environmental threat. Combined with the reaction with ozone to water vapour, hydrogen diffusion through the atmosphere beyond the stratosphere creates one of the strongest greenhouse gases on earth. By the production of harmful water vapour in this layer of the atmosphere, the hydrogen economy required for fuel cells would cause more environmental harm than benefits [39].

To further develop hydrogen technologies, including its generation, transportation, storage and use, i.e. in fuel cells, highly competent scholars and researchers with academic education are required. Consequently, the introduction of research and development, as well as educational facilities, could create spill-over effects in terms of a higher education of local residents. Like onshore power, fuel cell ships will only be sustainable if the hydrogen is produced by using renewable energy technologies. If not, fuel cell ships would be a zero-sum technology as well, shifting emissions from the ships and the port to another place. Last, a socio-environmental benefit of fuel cells is a reduction in sound pollution since fuel cells operate quietly and without vibrations. All taken together, people living near the port will enjoy improved living and working conditions [19].

11.3.3.3 Fuel switching

Fuel switching comprises the removal of impurities like nitrogen and sulphur, thus reducing NO_X and SO_X emissions during combustion. Nevertheless, these distilled fuels are still based on carbon and consequently transform to compounds like CO and CO_2 when burned. Hence, fuel switching will not induce the local benefits described above to the same extent as onshore power and fuel cells. Furthermore, fuel switching technology does not reduce the dependence on fossil fuels. Last, contrary to the other technologies, the sole application of fuel switching does not tackle noise pollution inflicted by auxiliary engines, as ships remain operating these very machines while docked in the port [23].

Table 11.2: Differences between expert and stakeholder analysis.

Technology	Onshore Power			Fuel Cell Ships			Fuel Switching		
Decision Criteria	AUS	WF	AWS	AUS	WF	AWS	AUS	WF	AWS
Economic profitability	2.5	3	7.5	2	3	6	2.75	3	8.25
Sustainability	2.5	2.5	6.25	2.5	2.5	6.25	2	2.5	5
Renewable energy supply	3.25	2.5	8.125	3.5	2.5	8.75	0	2.5	0
Compatibility	2.75	2	5.5	2	2	4	4.5	2	9
Simplicity	2.5	1	2.5	2	1	2	4.5	1	4.5
RESULTS			29.875			27			26.75

AUS = averaged unweighted score; WF = weighting factor; AWS = averaged weighted score. Weighting factor: 1 = less important, 2 = important, 3 = very important Score: 0 = not applicable at all, 1 = not applicable, 2 = not quite applicable, 3 = quite applicable, 4 = applicable, 5 = highly applicable

11.3.4 Energy technology portfolio and implementation planning

For the purpose of designing a technology portfolio, a decision matrix can now be made, following the method explained in Section 11.2 and based on the discussion of the technologies above (see Table 11.2).

Based on the range of possible ratings and weighting factors, the maximum possible score is 55 ($5 \times 3 + 5 \times 2.5 + 5 \times 2.5 + 5 \times 2 + 5 \times 1 = 55$). As displayed in Table 11.2, none of the three perceived technologies comes even close to this score, for several reasons. First, they are still in a more or less early stage of development and, thus, not yet mature in terms of economics and experience in implementation and operation. Second, depending on whether the technology has adaptive or disruptive effects on the existing energy system, the score for compatibility and simplicity will be either high or low, respectively. In any case, all three technologies still need further improvement in view of the outcomes of the decision matrix. Another remarkable detail concerns sustainability. Although three options should ensure a more sustainable port operation – they score moderately in this respect.

Onshore power attained the highest aggregate rating with 29.875 points due to its average performance in all domains. On the one hand, it comprises potential for the supply of renewable energy, given that the local utility shifts their energy production in such a trajectory. As it is unlikely to expect a predominantly renewable energy provision of such utilities in the near future, onshore power received an AUS of 3.25 [15]. Hence, onshore power not only imposes substantial costs but also requires considerable modifications and research into the implications of increasing power demand and fluctuations. The mediocre score on sustainability is vindicated by its tremendous socio-environmental impacts as elucidated in Section 11.3.3.

Based on the outcomes of Table 11.2, our technology portfolio will mainly consist of onshore power due to its relatively high scores in all domains. A broad implementation of fuel switching also makes sense as it is easily deployable and compatible with the existing energy infrastructure. However, fuel switching does not resolve concerns about fossil fuel dependence, so this technology might not be a solution for the far future. A technology that could be a solution, in the long run, is fuel cells. Due to restraints regarding infrastructure and huge costs, we do not expect this technology to be competitive in the near future. Also, fuel cell technology is still in an early development stage, requiring further research and development. Nevertheless, we think this technology has considerable potential, and for this reason, we include it in the stakeholder analysis and our recommendations.

11.4 Stakeholder analysis for the Port of Nansha

Our analysis up to now resulted in a technology portfolio for policymakers, decision-makers and other stakeholders who strive for more sustainable and responsible port activities. However, the feasibility of the actual implementation of these technologies depends on the social, institutional and other characteristics of a particular context. A stakeholder analysis helps to gain a greater understanding of such particularities and that is precisely what we will do for the Port of Nansha and district in the following section.

11.4.1 The necessity for the analysis of stakeholders and their values

Stakeholders have values, and various stakeholders may have conflicting values. Like conflicts of interests, although at a more fundamental level, such value conflicts may jeopardize the success of an intervention or even result in its failure. If they feel affected in their interests and values, local residents can, for example, unite and express their disagreement against local authorities and companies. Besides the earlier mentioned example of Shatian, there are myriad other cases of local outbursts. Villagers in Dongyang, Zhejiang, for example, protested against chemical plants causing severe environmental pollution by their industrial waste. Consequently, some plants had to be re-allocated and were even closed [40]. Besides the seizure of land by trespassing of property rights or otherwise [41], environmental disruptions may cause discontent [42].

Every group involved can resist. The Chinese government, for example, plans to ban gasoline-powered vehicles and works on timetables concerning the abolition of their production and sale. Alternatives like electric vehicles are supported in the form of subsidies and other policies, although the implications for oil producers might be disastrous [43]. Oil producers oppose this development. It would be better to keep these oil producers, who are also powerful stakeholders, somehow on board! To avoid or mitigate undesired effects, considering stakeholders from the government down to the grassroots of society is important. Since our technology portfolio and its implementation could cause socio-environmental damage to the local community, the significance of the inclusion of its values is indisputable.

11.4.2 Identification of stakeholders

Table 11.3 summarizes our findings of stakeholders in the Port of Nansha. The table consists of two parts. The upper part shows stakeholders influenced by and involved in any of the considered technologies, while the stakeholders in the lower part are solely involved in the development of particular technologies.

11.4.2.1 Onshore power

Concrete data was not found about actual manufacturers and producers of onshore power accessories, but merely about firms engaged in power system design and architecture, such as the Guangzhou Electric Power Design Institute [44]. They offer services ranging from consultancy to project building, extension and reconstruction. Another stakeholder in the manufacturer/producer section is Eaton. The firm specializes in power management (distribution and circuit protection) and backup power protection. The Guangzhou Power Supply Bureau is also an important stakeholder in designing, constructing and maintaining power grids. Guangzhou Sukai Electric Services, which aims to provide electricity directly to the consumers without the involvement of the national grid, is another private electricity retailer.

11.4.2.2 Fuel cell ships

The manufacturers of fuel cells are listed by their market share, with Dalian XYDL comprising 45% of the market, Jiangsu Huayuan 24%, Shanghai Shen-Li 12% and Shanghai Boneng 8%. Their customer base consists almost exclusively of scientists, teachers and academic personnel, with industrial customers that are only involved in demonstration projects for public transport, but not ships [22]. For the assessment of suppliers in the fuel cell industry, only fuel suppliers are included. Fuel cells suitable for maritime transportation either run on hydrogen or methane [19]. Hydrogen producing companies

Table 11.3: Stakeholders involved in the three technologies

General Stakeholders (Involved in Every Technology)			
Public Authorities	Provincial Government of Guangdong, Guangzhou Port Authority		
Port Operator	Guangzhou Port Group		
Residents	Local Citizens, Local Villagers		
Local Infrastructure	Local Schools, Hospitals, . . .		
Specific Stakeholders (Involved in Particular Technologies)			
	Onshore Power	Fuel Cell Ships	Fuel Switching
Manufacturers/ Producers	Guangzhou Electric Power Design Institute, Eaton	Dalian Xinyuan Dongli (Dalian XYDL), Jiang su Huanyuan, Shanghai Shen-Li, Shanghai Boneng . . .	CNPC, China National Offshore Oil Company (CNOOC), Sinopec
Suppliers (e.g. fuel)	Guangzhou Power Supply Bureau of China Southern Power Grid, Guangzhou Sukai Electric Services	**Foreign Companies in China** AirLiquide S.A., Linde Group . . . **Chinese Companies** China National Petroleum Corp. (CNPC), Sinopec	CNPC, CNOOC, Sinopec
Operator (e.g. National Grid, Ship Operators)	Ghuangzhou Power Supply Bureau of China Southern Power Grid	Non-Existent (merely demonstration and pilot plants)	Ships docking at Nansha Port (e.g. by MAERSK, etc.)

are either foreign, like AirLiquide from France and Linde Group from Germany [45, 46], or originate from China. The projects by AirLiquide and Linde, however, do not address fuel cells for maritime transportation, but other hydrogen-related ventures. As for Chinese suppliers, two of the biggest oil and natural gas producers, CNPC and Sinopec, superficially take into account hydrogen generation and distribution. While CNPC mentions hydrogen production only sporadically in their Corporate Social Responsibility Report [47], Sinopec lists hydrogen generation projects throughout China without mentioning the utility of the produced hydrogen [48]. Therefore, it is unclear to what extent Sinopec's projects contribute to the expansion of a hydrogen infrastructure specifically for fuel cells.

11.4.2.3 Fuel switching

Since this technology merely comprises the transition from heavy fuel oil to light, distilled fuel, the focus here is mostly on stakeholders from oil and gas companies in China. For this technology, no difference has to be made between producers and suppliers because oil and gas companies like CNPC and Sinopec either possess their own refineries or supervise external refinery companies [49, 50]. Like China National Offshore Corporation (CNOOC) – another major oil and gas company – CNPC and Sinopec also strive to improve the quality of their fuels to meet international standards [47, 51]. Besides them, ship operators are also important stakeholders for this technology. But since thousands of ships dock in the Port of Nansha every month and year, we decided to treat these stakeholders as one group.

11.4.3 The economic impact of projects on stakeholders

The deployment of the analysed technologies entails a host of economic, environmental and social implications. Previously, the socio-environmental impact has been dealt with generally, but it applies to the Port of Nansha as well. Consequently, in the following section, the technologies of onshore power, fuel cell ships and fuel switching are only assessed on their economic impact on the local stakeholders, specifying the earlier general considerations.

11.4.3.1 Onshore power

The return of investment for onshore power deployed in Copenhagen's port terminals comprised around 12 years [52]. For China, it might even take longer to obtain net profits due to differences in power quality and supply capacities, resulting in higher infrastructure expenses by public stakeholders like the Guangzhou Port Authority and the Provincial Government of Guangdong [53]. Stakeholders from the maritime transportation sector like ship operators are affected as well. Electricity coming from the shore tends to be more expensive than the fuel used for the auxiliary engines onboard, which has discouraging effects on profitability [29]. Other factors include retrofit costs for old ships and extra costs connected to the manufacturing of new ships with an adequate electrical management system. Furthermore, the construction of power infrastructure might also interfere with agricultural activities of farmers in the Nansha district. By reducing or even seizing their means of subsistence for the construction of grid extensions, these stakeholders might be affected by unemployment, lower incomes and bad prospects. However, the construction of new grid extensions as well as their operation and maintenance could also create jobs. Another benefit could be that compliance with the standards set by the IMO for emissions would prevent penalty payments that stakeholders like the Guangzhou Port Group would have to pay if they do not have this technology.

11.4.3.2 Fuel cell ships

If fuel cells should ever break their chains and become widely deployed, both ship operators and port operators, in our case the Guangzhou Port Group, would face monumental capital investment costs. With regard to the ships, major technical modifications are necessary, as fuel cell-driven ships also require adequate storage devices, power components and electric motors [21]. Port operators would have to establish a thorough hydrogen infrastructure with stable supply and distribution chains [54]. Hence, it is questionable, if and when these capital costs would be recovered. Stakeholders from oil and natural gas companies like CNPC, CNOOC and Sinopec might also be affected by fuel cell-powered ships. Due to the replacement of fossil fuel by hydrogen or methane, these companies will experience lower profits unless countermeasures like specialization in natural gas production are undertaken. Another option would be the diversification into hydrogen production, although this seems unlikely in the near future considering how superficially hydrogen is treated in their social responsibility reports [47, 48, 55].

Stakeholders of the fuel cell industry like Dalian XYDL and Jiangsu Huayuan would appreciate the broad expansion of their technology. With economic mechanisms such as learning and economies of scale, cost reductions could be achieved, which would consolidate the competitiveness of fuel cells and accessories [56]. Assuming broad implementation of fuel cell ships, local residents could also benefit by surging employment in the fuel cell industry and the hydrogen economy. Consequently, the incomes and standards of living of these stakeholders would be improved. However, these benefits in terms of employment could be offset by the loss of jobs in the incumbent energy system. While the expansion of fuel cell technology creates new business opportunities and jobs, existing companies, such as combustion engine manufacturers, might perish under the pressure of emerging competitors, causing redundancies.

11.4.3.3 Fuel switching

As institutions like MARPOL demand higher quality fuels with less sulphur, nitrogen and other substances, refineries will bear the consequences of fuel switching the most. Higher production costs reduce the profitability of refineries, operated by companies like CNPC. If a refinery is specialized in the wrong kind of distillate, prohibited by international standards, this would result in huge reductions in returns. Profits losses might be offset by the dismissal of personnel, resulting in higher unemployment. If Sinopec was affected, this would hit the province of Guangdong directly, since this company is active there as Sinopec Guangdong Oil Products Company [57]. However, if refineries do produce fuel complying with international standards, they will experience a huge increase in profitability [33]. In the meantime, ship operators will be obliged to purchase more expensive fuels, which will affect their balance sheets if the excess fuel costs are not compensated by subsidies such as in the case of the Port of Shenzhen [58, 59]. Compared with fuel cells, however, less capital will be required to adapt to the new technology. In the case of fuel switching, the infrastructure is already in place and well-known for decades, and ship modifications would only initially entail high investment costs [32, 33].

11.4.4 Identification of stakeholder values

11.4.4.1 General stakeholders

With regard to the general stakeholders of Table 11.3, the values could be taken from the paper by Ravesteijn, He and Chen [3], as their work refers to the same area as in this chapter, namely the Port of Nansha. The stakeholders and their values are listed as follows:

- Provincial Government of Guangdong: higher taxes/government revenue, higher GDP (gross domestic product), higher employment, less energy consumption
- Guangzhou Port Authority: higher throughput, more containers
- Guangzhou Port Group: more land for industry, profitability
- Local citizens: better air and water quality, less noise, better natural scenery
- Local villagers: better welfare, employment
- Local schools and hospitals: better air and water quality, less noise, better environment

11.4.4.2 Specific stakeholders

For the second type of stakeholders, the values are gathered from corporate reports and other publications. As the list below shows, more and more companies such as CNOOC and CNPC accept the principles of Responsible Innovation and Sustainable Development and determine their goals and commitments accordingly [47, 55]. Stakeholders of innovative technologies such as fuel cells, however, mainly aim for higher profits and the consolidation of their business through a broader customer base and infrastructure [22]:

- Onshore power facility producers: sustainable construction, better infrastructure, public welfare, knowledge exchange
- Fuel cell producers: higher profitability, broader customer base, more infrastructure

- Oil and natural gas companies (hydrogen and light fuel producers/suppliers):
  - CNOOC – sustainable growth, compliance with law and ethics, environmental responsibility, public welfare, stakeholder appreciation (shareholders, employees, etc.)
  - CNPC – sustainable growth, industrial growth/spill-over effects, societal welfare, shareholder appreciation, responsible operation (safety, environment, supply chain)

11.4.5 Identification of value conflicts

Public entities like the Provincial Government of Guangdong and the Guangzhou Port Authority prioritize economic values such as a higher GDP. Corporate stakeholders also endorse economic aspects, including profitability, but increasingly realize the significance of sustainability and social benefits of their actions. In contrast, private stakeholders like local residents are more concerned about environmental and social criteria, including employment. Hence, there is a potential for conflicts between stakeholders.

One of the conflicts is profitability versus socio-environmental responsibility. None of the analysed technologies increases the profitability of the stakeholders deploying it. Either high capital investment costs, as in the case of onshore power and fuel cells or operational costs like distilled fuels reduce the returns of corporations. Due to lower tax payments by these firms, local public authorities are also affected. As Ravesteijn, He and Chen [3] pointed out, clean air and water resulting from sustainable technologies are non-exclusive goods. Therefore, the implications of free riding – enjoying the benefits of investments without paying for them – are a considerable obstruction. However, clean air and water, less noise and undamaged nature are not only desirable from a flora and fauna perspective but also benefit the residents near the port. Thus, the local community is inclined to sacrifice the profitability of private companies for the sake of better health, more recreational opportunities and many other factors for improving life. This is contrary to the values of stakeholders like the Guangzhou Port Group, who strive for higher profitability and more land for industrial activity [3].

Due to a lack of available sources, there is intense competition between stakeholders, which creates a dissonance of accessibility versus liveability. While stakeholders like the Provincial Government of Guangdong claim land for industrial activities, others like local villagers demand agricultural plots to ensure their subsistence [3]. Therefore, the seizure of agricultural land or forest areas by project promotors is seriously undermining the lives of residents. Once the infrastructure for the supply and distribution of energy is established, be it onshore power or hydrogen technologies, the land in question is made inaccessible for local villagers and citizens. This results in a reduction of agricultural production, deteriorated mobility or even the loss of homes.

Based on these value conflicts, decision-making stakeholders should consider whether they are prepared to lose economic profit for socio-environmental benefits and to jump ahead as a paradigm of sustainable and responsible development.

11.5 Constitution of governance mechanisms for value conflict resolution

11.5.1 Stakeholder power and influence assessment

Stakeholders differ in power and influence. Following the paper by Ravesteijn, He and Chen [3], these differences will be assessed here. The following list ranks the power and influence of the stakeholders involved in the implementation of any technology from our portfolio, from the strongest to the weakest.

- Provincial Government of Guangdong
- Guangzhou Port Authority and Guangzhou Port Group
- Local citizens, villagers, schools and hospitals

The Provincial Government of Guangdong ranks in first position as it has the formal power to either start or stop a development project in the Port of Nansha. Guangzhou Port Authority and Guangzhou Port Group come next, holding ‘consultative power’ [3]. They can exert some influence, albeit without formal power. In contrast, local citizens, villagers, schools and hospitals usually have power and influence, for example, through the media or personal networks, but only a little. Nonetheless, these stakeholders can become a force to be reckoned with as soon as they unify their voices against unjust government and corporate actions.

For our stakeholder power and influence assessment, we assume that entities involved in specific technologies of our portfolio consist of manufacturers/producers, suppliers and operators. We treat these different types of stakeholders as equivalents of Ravesteijn, He and Chen’s [3] ‘Manufacturing Companies,’ involved in, among others, consultancy services. The stakeholders of each technology have a varying degree of power and influence. Fuel switching is the least disruptive technology with many stakeholders adhering to the incumbent energy system. Relatively, these stakeholders have the highest degree of power and influence as they act in favour of the existing infrastructure. Next are the stakeholders of onshore power. On the shore-side, onshore power heavily relies on the electricity supply and power grid utilities and involved stakeholders are influential. This does not apply to stakeholders from the ships. The stakeholders of fuel cell ships find themselves in the lowest positions. As the expansion of this technology would cause the disruption or destruction of the existing energy system, incumbent stakeholders will be inclined to suppress their competitors by keeping their impact low. Furthermore, the network of fuel cell stakeholders is still porous, with a limited customer base and a low number of producers.

11.5.2 Policy recommendations based on stakeholder groups

Equipped with a technology portfolio and some understanding of the social context of Nansha district, especially regarding its stakeholders and value conflicts between them, the final step of our stakeholder analysis can be taken: policy recommendations for the implementation of our technology portfolio considering stakeholder values and conflicts (Table 11.4).

Two remarks in advance. First, we divide the stakeholders affected by the recommended policies in the groups suggested by Geels [13]. Here, we do not exclude any stakeholder groups. By combining producers and suppliers, we focus on six stakeholder groups:

- Producers and suppliers
- Financial networks
- Research networks
- Public authorities
- Societal groups
- Consumers/residents

Second, it is important to note that our policy recommendations are based on our analysis and understanding thus far. We use some additional literature, but not when it comes to past and current energy policies in China (see for this topic [60–65]).

We will address profitability versus socio-environmental responsibility by ensuring the profitability of stakeholders who will act socio-environmentally responsibly, while accessibility versus liveability is tackled by compensating stakeholders whose life quality will be affected by the implementation of the technology portfolio. In addition, we address the empowerment of currently weak stakeholders.

11.5.2.1 Producers and suppliers

Some of the reasons why stakeholders from this group are reluctant to engage in innovative technologies are the high uncertainty and risk connected to them [66]. This applies to entrepreneurs from both start-ups and established companies trying to diversify their business, although the former are more vulnerable to these obstacles than the latter [67]. These problems could be addressed by reducing the risk of investing in innovative technologies, which could be achieved by establishing an innovation fund and frameworks for risk management. Another possibility would be tax advantages for initial buyers of innovative products, thus broadening the customer base of these stakeholders. With such measures, producers and suppliers could operate more profitably, while developing socio-environmentally responsible products.

11.5.2.2 Public authorities

Public support in the form of subsidies and favourable tax regimes is a good way to ensure the profitability of socio-environmentally engaged companies. While these policies would support them directly by providing capital, other policies like energy and/or pollution caps would incentivize their innovative technologies indirectly by restraining the use of energy- and emission-intensive products [15].

In addition, political reforms are required that tackle the issue of corruption and that could bring about a more democratic political constitution [68]. In addition to political freedom, reforms should also promote further economic liberalization [14]. Such policies would give new impetus to the resolution of value conflicts regarding profitability and liveability in the Port of Nansha and district because a more open and democratic political-economic system would improve the position of local residents in public decision-making, making them better able to stand up for their property and other rights.

11.5.2.3 Societal groups

Democratic reforms would not only enable local stakeholders to raise their voice, but also to unite and organize in societal groups (non-governmental groups). As such, they would represent the will of the people and be able to push regional and national policymaking in desired directions, all through peaceful activism and lobbying. They could oppose and expose stakeholders acting in a socio-environmentally harmful way and raise public awareness about their malpractices. In the Port of Nansha area, such social groups can appeal to stakeholders such as the Guangzhou Port Authority and the Provincial Government of Guangdong and denounce the irresponsible technologies currently used in the port. This would contribute to reconciling the values of accessibility and liveability, as property rights of less powerful stakeholders would be better protected, while industrial conduct would be monitored.

11.5.2.4 Consumers/residents

Product labelling could raise consumer awareness. By pointing out the sustainable and responsible background of a product, including transit through the port, the Port of Nansha in this case, customers may be inclined to consider that in their buying behaviour. This indirectly addresses the profitability versus socio-environmental responsibility conflict. If customers would prefer more

Table 11.4: Policy recommendations tackling stakeholder value conflicts

	Profitability vs. Socio-Environmental Responsibility	Accessibility vs. Liveability
Producers/Suppliers	Innovation Fund Risk Management Tax Advantages	n.a.
Public Authorities	Subsidies Favourable Tax Regimes Energy and Pollution Cap	Political Reform
Societal Groups	n.a.	Formation and Liberalization of Societal Groups
Consumers/ Residents	Product Labelling Information Centre	Compensation Fund

sustainable and responsible products, the Guangzhou Port Group might be induced to deploy technologies that both meet customer needs and maintain their revenues.

Another suggestion is the setting up of information centres to raise awareness about renewable and conventional energy technologies among customers and other stakeholders. By providing education, people form their own opinion and are able to express their values and desires accordingly. Again, this policy would have indirect effects, but this time on the accessibility versus liveability opposition. A well-informed community is more likely to resist socio-environmentally harmful technologies and to do so based on facts and hard data. Last, a local government compensation fund could directly support stakeholders who lose their subsistence or even home due to energy technology implementation. If residents know that they will receive appropriate compensation in the form of money or a new rental contract in another area, value conflicts like accessibility versus liveability might be mitigated, if not satisfactorily resolved.

11.6 Conclusion: Results and prospects

11.6.1 Research results

First, we will address the sub-questions, followed by an answer to the main research question in the next section.

1. What energy technologies are available to contribute to a more sustainable and responsible operation of the Port of Nansha and what are their economic and socio-environmental characteristics?

   Three technologies have been chosen because of their relevance to sustainable and responsible port operations: onshore power, fuel cells and fuel switching. All of these can reduce emissions of pollutants in the Port of Nansha, although to varying degrees. While onshore power and fuel cells will not cause any hazardous emissions at all, fuel switching will merely mitigate the exhaustion of fume gases due to fuel impurities. Carbon-based emissions like CO_2, essentially, will not be lowered. Onshore power, fuel cells and fuel switching have different fixed and variable costs depending on the level of innovation. Technologies that adapt to the incumbent energy system such as fuel switching have lower costs than disruptive technologies such as fuel cells, which upends existing supply and distribution chains. Socio-environmental impact varies as well. Options that require new infrastructure like onshore power and fuel cells lead to huge environmental disruptions due to construction activities.
2. How can a portfolio of available energy technologies be composed on the basis of predefined criteria and what are the timescales for the individual technologies to be deployed in the most feasible fashion?

   After the assessment, the three technologies have been rated and evaluated in a comparative fashion by a decision matrix. Applying the five criteria of economic profitability, sustainability, renewable energy supply, compatibility and simplicity, onshore power appeared to be the most suitable option for a more sustainable and responsible operation at the Port of Nansha, followed by fuel switching. Both technologies are eligible for short- and medium-term implementation, although the dependency on fossil fuels ensures that fuel switching is not feasible for the distant future. Therefore, fuel cells are recommended for far-future implementation since their extensive requirements in infrastructure and cost reduction impose insurmountable barriers to near-future deployment.
3. Who are the local stakeholders affected by the economic, environmental and social implications of technology implementation and what are the possible value conflicts among these stakeholders?

   The relevant stakeholders in the Port of Nansha and district include regional and local authorities, as well as local citizens and villagers. There are also stakeholders such as manufacturers, suppliers and operators, varying according to the technology. These stakeholders are involved in and affected by the implementation of the technologies by shifts in local labour and commodity markets as well as changing market shares in the energy sector. Regarding the socio-environmental impact, the proposed technologies and their deployment can interfere with the subsistence of local residents. Public stakeholders and companies appeared to recognize the significance of sustainability, although profitability still prevails. In contrast, private stakeholders like citizens and villagers are more focused on a clean environment and social values. Based on these stakeholder values, two value conflicts have been detected, namely profitability versus socio-environmental responsibility and accessibility versus liveability.

4. Which policies and institutions could reconcile stakeholder value conflicts to improve the feasibility of the technology portfolio implementation?

   A set of policies and recommendations have been designed to mitigate the identified value conflicts. These aim to improve the profitability of socio-environmental responsible stakeholders, thus supporting their endeavours to pitch their business ideas. Such policies include, among others, innovation funds, tax advantages and favourable credit schemes. Additionally, policies such as political reforms, formation and liberalization of societal groups and compensation funds can tackle the accessibility versus liveability opposition by strengthening property rights, on the one hand, and improving the performance and quality of energy products, on the other hand. In this respect, public authorities at all levels can support local stakeholders through subsidies and favourable tax regimes for both producers and consumers.

11.6.2 Final conclusion

How can sustainable and responsible development of the Port of Nansha be advanced through innovative energy technologies and which policies and institutions could support the implementation of such technologies in view of local stakeholders and their values?

Energy technologies, such as onshore power, fuel cells and fuel switching, can contribute considerably to a more sustainable and responsible development of the Port of Nansha. Socio-environmental benefits include, among others, reduced levels of local emissions and less noise. Consequently, not only the labour conditions of ship and port personnel will be improved, but also the life quality of local residents near the port. These technologies can entail job opportunities, i.e. in grid operation and maintenance or in the fuel cell industry. Undesired effects range from environmental disruptions due to vast construction sites to the undermining of incumbent energy systems by innovative technologies like fuel cells. Hence, the implementation of individual technologies needs to be planned carefully and in such timescales that the affected stakeholders have a chance to prepare themselves. Since the values of the involved stakeholders can vary considerably, decision- and policymakers need to include them and offer them support in the form of appropriate policy interventions to resolve and mitigate value conflicts. Political reforms that, among others, consolidate and ensure property and other rights seem almost inevitable in China if it wants to continue its growth. Furthermore, government support in the form of subsidies and tax incentives is also needed to encourage entrepreneurs to make a leap towards greater sustainability and responsibility.

11.6.3 Recommendations for further research

In retrospect, many recommendations for further research result not only from knowledge gaps in this chapter but also from the limitations of our methodology. First, only three technologies were analysed in depth, but there are myriad other technologies such as carbon capture and storage (CCS) and microgrids, which could contribute to more sustainable and responsible port operations. Second, the policies listed are merely qualitative suggestions, the efficiency and effectiveness of which have remained unproven. It would be interesting to learn more about their effects on society, for example, by agent-based modelling (ABM). Last, we strongly advocate the inclusion of local stakeholders in studies like this. The authors of this contribution do not have a connection with China and its culture. Involving stakeholders from different domains and power levels in the determination of a suitable technology portfolio would strongly improve the quality of such work. This can be achieved through extensive field studies and interviews with stakeholders on the ground. With these recommendations in mind, the Port of Nansha could be a step closer to improved operation of the port and the idea of establishing a paradigm of Responsible Port Innovation becomes closer to reality rather than wishful thinking.

Acknowledgements

This chapter is based on a paper the authors successfully concluded on the Technology and Global Development course at Delft University of Technology. Wim Ravesteijn, the module manager of the course, thoroughly edited the text.

References

[1] Fallon, T., The New Silk Road: Xi Jinping's grand strategy for Eurasia. *American Foreign Policy Interests*, 37(3), pp. 140–147, 2015.

[2] Bhattacharyay, B.N. & Prabir, D., Restoring the Asian Silk Route: Toward an integrated Asia. Tokyo: Asian Development Bank Institute (ADBI), 2009. http://hdl.handle.net/10419/53649. Accessed on: 15 March 2018.

[3] Ravesteijn, W., He, J. & Chen, C., Responsible innovation and stakeholder management in infrastructures: The Nansha Port Railway Project. *Ocean & Coastal Management*, 100, pp. 1–9, 2014.

[4] Guangzhou Port Authority, About Us. http://www.gzport.gov.cn/portal/site/site/portal/english/showContent.jsp?contentId=TX94G0MZHD34C5Z4NH87TC7OQD27474&categoryId=AG3BUV9GW5XDYDB157UTJW3GD6442B6K. Accessed on: 15 March 2018.

[5] Boer de, C., Hogeveen, M. et al., Sustainable Port Development: Port of Guangzhou, China. Delft: Delft Infrastructure & Mobility Initiative, 2013.

[6] Wang, L. et al., Functional differentiation and sustainability: A new stage of the Chinese container port system. *Sustainability*, 9(3), p. 328, 2017.

[7] Ng, S.K.W., Controlling Ship Emissions in the Pearl River Delta. Hong Kong: Civic Exchange, 2016.

[8] West of England, China – Emission Control Areas, 2015. https://www.westpandi.com/publications/news/china-emission-control-areas/. Accessed on: 19 May 2021.

[9] Fung, F. et al., Prevention and Control of Shipping and Port Air Emissions in China. Natural Resources Defence Council, Beijing, 2014.

[10] China, Vision for Maritime Cooperation under the Belt and Road Initiative, 2017.

[11] Watt, A., Project Management, The Saylor Foundation, 2016. http://www.opentextbooks.org.hk/system/files/export/15/15694/pdf/Project_Management_15694.pdf. Accessed on: 19 May 2021.

[12] Alphen van, K., Sark van, W.F.J.H.M. & Hekkert, M.P., Renewable energy technologies in the Maldives: Determining the potential. *Renewable and Sustainable Energy Reviews*, 11, pp. 1650–1674, 2007.

[13] Geels, F.W., Technological transitions as evolutionary reconfiguration processes: A multi-level perspective and a case study. *Research Policy*, 31, pp. 1257–1274, 2002.

[14] Acemoglu, D. & Robinson, J.A., *Why Nations Fail: The Origins of Power, Prosperity, and Poverty*. New York: Crown Publishers, 2012.

[15] Wang, H., Mao, X. & Rutherford, D., Costs and Benefits of Shore Power at the Port of Shenzhen. International Council on Clean Transportation, Beijing, 2015 (December).

[16] Yang, C.-S., Evaluating the use of alternative maritime power in Taiwan. *Maritime Business Review*, 1(3), pp. 208–224, 2016.

[17] Arduino, G., Carrillo Murillo G.D. & Ferrari, C., Key factors and barriers to the adoption of cold ironing in Europe, Working Papers 11_15, SIET

Società Italiana di Economia dei Trasporti e della Logistica, revised 2011.

[18] Alaswad, A. et al., Developments in fuel cell technologies in the transportation sector. *International Journal of Hydrogen Energy*, 41, pp. 16499–16508, 2016.

[19] Biert van, L. et al., A review of fuel cell systems for maritime applications. *Journal of Power Sources*, 327, pp. 345–364, 2016.

[20] de-Troya, J.J. et al., Analysing the possibilities of using fuel cells in ships. *International Journal of Hydrogen Energy*, 41, pp. 2853–2866, 2016.

[21] Winkler, W., *Encyclopedia of Electrochemical Power Sources*. Oxford: Elsevier, 2009.

[22] China Strategic Research, China's Fuel Cells Market is a Long Way Off, http://www.gcis.com.cn/china-insights/industry-articles/143-china-s-fuel-cells-market-is-a-long-way-off, n.d. Accessed on 15 March 2018.

[23] Abadie, L.M., Goicoechea, N. & Galarraga, I., Adapting the shipping sector to stricter emissions regulations: Fuel switching or installing a scrubber. *Transportation Research* Part D, 57, pp. 237–250, 2017.

[24] Lack, D.A. et al., Impact of fuel quality regulation and speed reductions on shipping emissions: Implications for climate and air quality. *Environmental Science & Technology. ACS Publications*, 45(20), pp. 9052–9060, 2011.

[25] Mestl, T. et al., The doubtful environmental benefit of reduced maximum sulphur limit in international shipping fuel. *Environmental Science & Technology. ACS Publications*, 47(12), pp. 6098–6101, 2013.

[26] ABS, Fuel Switching Advisory 2015, 2015.

[27] Fiadomor, R., Assessment of alternative maritime power (cold ironing) and its impact on port management and operations, World Maritime University, 2009. https://commons.wmu.se/all_dissertations/277/. Accessed on: 15 March 2018.

[28] Theodoros, P.G., A Cold Ironing Study on Modern Ports, Implementation and Benefits Thriving for World-wide Ports, National Technical University of Athens, 2012. http://www.ops.wpci.nl/_images/_downloads/_original/1379920206_wmudissertation.pdf. Accessed on: 19 May 2021.

[29] Eastern Research Group and Energy & Environmental Research Associates, Shore Power Technology Assessment at U.S. Ports. US Environmental Protection Agency, Washington, DC, 2017. https://www.epa.gov/sites/production/files/2017-05/documents/420r17004-2017-update.pdf. Accessed on: 19 May 2021.

[30] Vogler, F. & Sattler, G., 3: Hydrogen-fuelled marine transportation. In: Ball, M., Basile, A. & Veziroğlu, T.N. (eds), *Compendium of Hydrogen Energy*. Oxford: Woodhead Publishing, pp. 35–65, 2016.

[31] Bunkerindex, Bunker Index Summary Page, 2017.

[32] Sulphur cuts will have some countries sensing profit, 2009. https://www.thenational.ae/business/sulphur-cuts-will-have-some-countries-sensing-profit-1.514524. Accessed on: 15 March 2018.

[33] George, L., Shipping fuel change to upend refining - and threaten OPEC crudes, 2016. https://www.reuters.com/article/us-oil-refineries-sulphur/shipping-fuel-change-to-upend-refining-and-threaten-opec-crudes-idUSKBN1321JE. Accessed on: 15 March 2018.

[34] Brauer, M. et al., Air pollution and the development of asthma, allergy and infections in a birth cohort. *European Respiratory Journal*, 29(5), pp. 879–888, 2007.

[35] Johansen, B., *Global Warming* 101, ABC-CLIO, 2008.

[36] Black, B. & Weisel, G.J., *Global Warming*, ABC-CLIO, 2010.

[37] Swim, J. et al., Psychology's contributions to understanding and addressing global climate change. *American Psychologist*, 66(4), pp. 241–250, 2011.

[38] Yang, J. et al., Effect of ferrite on the hydrogen embrittlement in quenched-partitioned-tempered low carbon steel. *Material Science & Engineering*, 712, pp. 630–636, 2018.

[39] Akitt, J.W., Some observations on the greenhouse effect at the Earth's surface. *Spectrochimica Acta Part A: Molecular and Biomolecular Spectroscopy*, 188, pp. 127–134, 2018.

[40] Griffith-Jones, S., Ocampo, J.A. & Spratt, S., Financing renewable energy in developing countries: Mechanisms and responsibilities, Mobilising European Research for Development Policies (February), 2012. http://academiccommons.columbia.edu/download/fedora_content/download/ac:153327/CONTENT/Financing_Renewable_Energy_in_Developing_Countries.pdf. Accessed on: 19 May 2021.

[41] Yep, R., Containing land grabs: A misguided response to rural conflicts over land. *Journal of Contemporary China*, 22(80), pp. 273–291, 2013.

[42] Ren, B. & Shou, H., *Chinese Environmental Governance: Dynamics, Challenges, and Prospects in a Changing Society*. New York: Palgrave Macmillan, 2013.

[43] LeSage, J., China's ban on gas-powered cars could cripple the oil market, http://www.businessinsider.com/china-ban-gas-powered-cars-effect-

on-oil-market-2017-9?international=true&r=US&IR=T, 2017. Accessed on 15 March 2018.

[44] Guangdong Electric Power Design Institute, n.d.

[45] Linde, H2 Reference Projects. http://www.linde-gas.com/en/innovations/hydrogen_energy_h2/reference_projects/index.html, n.d. Accessed on: 15 March 2018.

[46] AirLiquide, China: Strengthening in the hydrogen market. https://www.airliquide.com/media/china-strengthening-hydrogen-market, 2012. Accessed on: 15 March 2018.

[47] CNPC, Corporate Social Responsibility Report, 2015. http://csr.cnpc.com.cn/. Accessed on: 19 May 2021.

[48] Reuters, Sinopec to launch first green hydrogen project in 2022. https://www.reuters.com/business/sustainable-business/sinopec-launch-first-green-hydrogen-project-2022-2021-05-25/, 25 May 2021. Accessed on: 24 June 2021.

[49] Sinopec, Refining, http://www.sinopecgroup.com/group/en/business/Refining.shtml, n.d. Accessed on: 15 March 2018.

[50] Petrochina Refineries, http://abarrelfull.wikidot.com/petrochina-refineries, 2015. Accessed on: 15 March 2018.

[51] Sinopec starts supplying cleaner fuel to Beijing in anti-smog drive. http://www.hydrocarbonprocessing.com/news/2017/01/sinopec-starts-supplying-cleaner-fuel-to-beijing-in-anti-smog-drive, 2017. Accessed on: 15 March 2018.

[52] Ballini, F. & Bozzo, R., Air pollution from ships in ports: The socio-economic benefit of cold-ironing technology. *Research in Transportation Business & Management*, 17(Supplement C), pp. 92–98, 2015.

[53] Zhang, X., Analysis of the incentives in environmental strategies implementation in Chinese ports. Erasmus University of Rotterdam, Rotterdam, 2016.

[54] Moreno-Benito, M., Agnolucci, P. & Papageorgiou, L.G., Towards a sustainable hydrogen economy: Optimisation-based framework for hydrogen infrastructure development. *Computers and Chemical Engineering*, 102, pp. 110–127, 2017.

[55] CNOOC, Social Responsibility Report, 2013. http://www.cnoocltd.com/. Accessed on: 19 May 2021.

[56] Blok, K. & Nieuwlaar, E., *Introduction to Energy Analysis*. Oxford: Taylor & Francis, 2016.

[57] Sinopec, Sinopec Guangdong Oil Products Company. http://www.sinopec.com/listco/en/about_sinopec/subsidiaries/oil_products

_sales/20161109/news_20161109_390446318900.shtml, n.d. Accessed on 15 March 2018.

[58] Shenzhen Municipal Transportation Commission, Interim Measures for the Administration of Shenzhen Port, Ship Shore Facilities and Marine Low-Sulfur Oil Subsidy Funds, Shenzhen, 2014.

[59] Shenzhen Municipal Transportation Commission, Shenzhen Port, Ship Shore Power Facilities, and Marine Low Sulfur Oil Subsidies Implementation Details, Shenzhen, 2015.

[60] Li, M. & Patiño-Echeverri, D., Estimating benefits and costs of policies proposed in the 13th FYP to improve energy efficiency and reduce air emissions of China's electric power sector. *Energy Policy*, 111, pp. 222–234, 2017.

[61] Wu, G. et al., The elasticity of the potential of emission reduction to energy saving: Definition, measurement, and evidence from China. *Ecological Indicators*, 78, pp. 395–404, 2017.

[62] Tseng, S.-W., The Effectiveness of China's Energy Conservation Policy: On the Industry by the Revised Divisia Index Method. *ACM International Conference Proceeding Series*, F128273, pp. 25–29, 2017.

[63] Wang, Z., Wang, X. & Guo, D., Policy implications of the purchasing intentions towards energy-efficient appliances among China's urban residents: Do subsidies work? *Energy Policy*, 102, pp. 430–439, 2017.

[64] A, R.J. van der, Mijling, B., Ding, J., Elissavet Koukouli, M., Liu, F., Li, Q., Mao, H., & Theys, N., Cleaning up the air: Effectiveness of air quality policy for SO_2 and NOx emissions in China. *Atmospheric Chemistry and Physics*, 17, pp. 1775–1789, 2017.

[65] Karplus, V.J., Rausch, S. & Zhang, D., Energy caps: Alternative climate policy instruments for China? *Energy Economics*, 56, pp. 422–431, 2016.

[66] Teng, Y. et al., Analysis of stakeholder relationships in the industry chain of industrialized building in China. *Journal of Cleaner Production*, 152, pp. 387–398, 2017.

[67] Geels, F.W. & Schot, J., Typology of sociotechnical transition pathways. *Research Policy*, 36, pp. 399–417, 2007.

[68] Zhou, Z., Government ownership and exposure to political uncertainty: Evidence from China. *Journal of Banking and Finance*, 84, pp. 152–165, 2017.

CONCLUSION

The Dao of a sustainable and responsible 21st Century Maritime Silk Road

Wim Ravesteijn

Department of Values, Technology and Innovation, Delft University of Technology, The Netherlands

Beijing Information Science and Technology University, China

> The Dao of heaven benefits
> rather than harms all things;
> The Dao of the sage is to give
> rather than rob the people.
> Lao-tzu, Dao De Jing, 6th century BCE

C.1 Introduction

This chapter summarizes the main conclusions from the contributions to this book and looks forward to the future of the 21st Century Maritime Silk Road. Four main questions have guided and structured the work:

1. Are the port construction and expansion projects along the new Maritime Silk Road, both in China and elsewhere with Chinese investments and assistance, responsible and sustainable? And if not, how could activities be improved?
2. Are these projects promoting socio-economic development, especially in the 'host countries' concerned? And if not, how could that be achieved?
3. Is China, as the initiator of the 21st Century Silk Road programme and in line with its intentions as suggested in all policy documents, showing 'moral leadership' in responsible and sustainable port innovation and development?
4. Together with Sustainable Development, Responsible Innovation constitutes the perspective from which we look at things, but what can we learn from port practices and related developments in terms of this concept? In other words, what would Responsible Port Innovation look like?

These lead questions have not been addressed simultaneously in the various chapters; not even one chapter has dealt with all of them. However, all chapters discuss at least one of these questions, most chapters more than one. This is the place to bring all information together briefly and see what has been achieved in terms of these questions. Questions 1, 2 and 3 will be discussed in Sections C.2 and C.3. A separate section will be devoted to question 4. The final considerations in Section C.5 conclude this chapter and book.

C.2 Port innovation: current practices and possibilities for improvement

The contributions in Part I are about our research approach, the cultural backgrounds of the Silk Road area, the relevant Chinese policies and the lessons to be learned from eco-cities that set the stage for the case studies of port projects in Parts II and III. These cases show ecological sustainability considerations are often part of the game, although much has still to be implemented, especially in Asia and Africa. Onshore power appeared to be an attractive technology for reducing CO_2 and other emissions. The Chinese are enthusiastic about it, but it is not playing a major part in their partnerships with non-Chinese ports along the Silk Road. At home, they are confronted with increasing pressures to be sustainable, from politics and the people, but abroad this pressure is apparently lacking.

These contributions, however, also show the way forward towards responsible and sustainable port innovation and development, mostly through involving stakeholders and paying more attention to public values. International cooperation can support new initiatives, adopting best practices from elsewhere, including Chinese examples, to inspire construction activities in other places. Other possibilities concern consulting experts and using up-to-date technological and scientific sources, as much specialized knowledge is available regarding supportive planning tools, technologies and methods of technology choice, procedures of stakeholder participation and addressing conflicting values and so on, as the chapters in these parts show.

Generally, with its sustainable approach and its institutionalized stakeholder consultation, the Rotterdam Maasvlakte 2 Port seems to be one of the best examples of Responsible Port Innovation, although onshore power is slow to implement, and most energy is supplied by traditional power plants (see also [1]). Several Chinese ports, however, are also pioneers for a better future, as evidenced by the environmentally sound port innovation in Dalian and Shanghai.

C.3 Ports and development: impact and adjustment

With regard to economic growth and social development, our case studies have been much less conclusive, not least because the development effects of port projects have not been a central topic. Port plans in China fit in with the economic and social processes of transformation that take place in the country, and there is no doubt these plans will have positive impacts on the economy and society. The same might apply to port development in The Netherlands and maybe Vietnam and Greece, as required supportive institutional and political conditions are present.

However, some other countries show a different picture, which could be expected after the description, analysis and suggestions for improvement of the conditions in especially Asian countries in Part I. Promoting socio-economic development is usually an important goal of port projects, but this is not always how things work out. In Pakistan, port construction may result in economic growth, but full socio-economic development, another dimension of Sustainable Development, is not happening yet because of the kind of institutional and cultural challenges that have been discussed in the Introduction. The same applies to Tanzania. Both countries face an increased debt burden and increased dependence on China. Consequently, these countries could experience 'underdevelopment' [2] rather than – or in combination with – development, especially when it comes to specific social or ethnic groups.

C.4 Ports, innovation, development and leadership: Is China showing the way?

In terms of Responsible Port Innovation, China is definitely on the way forward and several cases have shown that new relevant technological practices are being applied (green in Dalian and Shanghai), while new approaches are being proposed as well, in terms of management (Harmony model) and scientifically (sustainable and responsible cost–benefit analysis). The Maasvlakte 2 port in Rotterdam may be top of the bill at the moment, but the Port of Shanghai is certainly on the rise (see also [3]).

A Chinese Responsible Port Innovation model might arise out of this, inspired by the experience with eco-cities, which could be followed by other countries. Although China is not so actively spreading this emerging model, in conformity with its practice and history of non-interventionism, it could take more 'moral leadership' to do so, which would not only be in line with its old philosophical traditions ('leading by giving a good example') but would also fit in with the new role the country aspires – and is expected – to play on the world stage.

If we broaden the picture to development in general, China is even less active in spreading its practices, up to the point that it might contribute to 'underdevelopment' rather than development, especially when it comes to indebtedness. Despite optimistic views ('the Chinese are our friends' [4]), the Chinese development model for other countries, which is ultimately based on its own successful experiences and which should provide an alternative for the

Western model, is not an approach that involves China's own best practices. To put it sharply, it is more of an approach that puts China's interests first and that leaves other countries free to be corrupt and to do things that are bad for them. In doing so, it looks as though China is avoiding the responsibilities it officially wants to have. Actually, this seems to be precisely the reason why some leaders embrace and welcome the Chinese and turn away from the 'paternalist' West, but this is not necessarily an improvement for the people of these countries.

C.5 Academic lessons: Responsible Port Innovation

Responsible Innovation, in its application to ports and within the context of Sustainable Development, has been used in several meanings or conceptualizations in this volume. Common characteristics are attention to the environment, inclusion of stakeholders and considering values. Several chapters addressed its key issue: how to abridge value differences or even oppositions. This resulted in a number of interesting methods, such as the Low Carbon Emission technology choice model and the Harmony model.

The Harmony model presupposes or emphasizes the significance of a comprehensive and collectivist and, consequently, qualitative approach, while the technology choice model provides for an analytic, detailed and as quantitative as possible approach. These methods enrich the toolbox of Responsible Innovation when it comes to dealing with different, competing or even conflicting values. Ultimately, the suitability of all these methods depends on the specific situation, which makes the search for 'local solutions' (diversity) the best method after all [5].

Bringing this argument one step further, what we can definitely learn from the contributions in this book is that there is no agreement about what 'a responsible innovation' would be. What we usually consider as such, is only the Western version of it, which is not automatically shared by actors in other countries. Such an innovation cannot be defined or determined separately from the social and cultural context, except in general and thus vague terms (like involving stakeholders and considering their values). Like Sustainable Development, Responsible Innovation not only develops over time, in response to changing mindsets and value systems, it also takes shape in ways that reflect people's traditions. While, for example, in the West, responsible and sustainable innovation and development provide much room for strategic negotiations between conflicting partners for the sake of an acceptable compromise, in China, a pursuit of harmony and consensus is much more visible.

Combined and applied to ports, Responsible Innovation and Sustainable Development merge into Responsible Port Innovation. Starting from the primacy of the local context, the various chapters have presented a large number of insights, methods and approaches that have given this concept form and content. Specific sustainable energy technologies and technological concepts are at stake, such as onshore electricity and port clusters, while bringing together the many direct and indirect stakeholders and their different values requires adapted procedures, such as the Harmony model. The peculiarity of the concept, however, is in the first place the diversity and specificity of the application, tailored to ports as a special kind of technological complex. This means that the 'mother concepts' of Responsible Innovation and Sustainable Development remain fully in place.

C.6 The future of the New Silk Road

Will the 21st Century Silk Road programme lead the world towards a prosperous, peaceful and environmentally sound future or will it pave the way for the collapse of civilization because of its disruptive consequences for society, the environment and the climate? Will it, ultimately, prove the classical economists were right in their expectation that economic growth is limited after all or will it be another proof that stationary economics do not exist?

The answers are in our hands. Classical economists like Thomas Malthus and David Ricardo were wrong in their gloomy expectations because technology developed beyond their imagination. Likewise, we can turn the New Silk Road into a success, in terms of infrastructure and its development effects, on the condition that we take all measures necessary to see to it that the additional growth it will most probably entail – in China, Asia, Africa, Europe and the West in general – will be turned into socially and environmentally sound development. This requires huge efforts, besides social reforms in various countries. The chapters in this book have shown this is not only possible but also already happening here and there, although counterproductive effects are also visible.

Whether the New Silk Road will also lead to a more peaceful world is hard to tell. Peace depends on prosperity, but there is a lot more to consider. More intense contacts and exchange between people could also lead to more tensions, especially if identity issues are at stake. Increasing Chinese influence in the world counters and mitigates Westernization to a certain extent. Consequently, it could result in greater harmony worldwide, but it could also induce and strengthen the 'clash of civilizations' foreseen by Samuel Huntington [6]. One contribution in Part I dealt with these issues for Asia, but its argument applies likewise to the world as a whole.

These issues are difficult to deal with, but there is one assumption that we might question, namely the continued existence of 'mainports' as engines of trade and development in the world. Advancements in 3D printing technologies might decrease the significance of transportation and, consequently, ports in the future [7]. This might be good news for the environment, but bad news for all those involved in the construction and management of ports and who directly or indirectly profit from these hubs. However, the Silk Road dates back at least some 2000 years, and a future without it seems unthinkable. This future could be bright for all involved if we act in the spirit of Lao-tzu, rephrasing his words as follows:

The Dao of the SILK ROAD benefits
rather than harms all things;
the Dao of the ENGINEER is to give
rather than rob the people.

References

[1] Port of Rotterdam, Maasvlakte 2, https://www.portofrotterdam.com/en/our-port/our-themes/a-sustainable-port/maasvlakte-2. Accessed on 24 June 2021.

[2] Frank, A.G., *The Development of Underdevelopment.* Boston, MA: New England Free Press, 1966.

[3] Geenhuizen, M. van, Song, L. & Ravesteijn, W., Seaport development and accelerating energy transition: Could Rotterdam and Shanghai take on a leadership role? In: Geenhuizen, M. van, Holbrook, J.A. & Taheri, M., *Cities and Sustainable Technology Transitions: Leadership, Innovation and Adoption.* Cheltenham: Edward Elgar Publishing, pp. 189–211, 2018.

[4] Moyo, D., *Dead Aid: Why Aid Makes Things Worse and How There Is Another Way for Africa.* London: Penguin, 2010.

[5] Poel, I. van de, Values in engineering design. In: Meijers, A., (ed.), *Philosophy of Technology and Engineering Sciences.* Amsterdam: Elsevier, pp. 973–1006, 2009.

[6] Huntington, S.P., *The Clash of Civilizations and the Remaking of World Order.* New York: Simon & Schuster, 1996.

[7] Planbureau voor de Leefomgevingen Infrastructuur, Mainports voorbij, The Hague, 2016.

Authors

- Aida Cahyaningwidi, MSc, Sustainable Energy Technology, Faculty of Technology, Policy and Management, Delft University of Technology, The Netherlands
- Larissa Eggers, MSc, Delft University of Technology, The Netherlands
- Oscar Garay Garcia, MSc, Delft University of Technology, The Netherlands; South America Energy Business Development, Schneider Electric Energy and Sustainability Services, Chile
- Andreas van Giezen, MSc, Delft University of Technology, The Netherlands
- Carola Hein, Professor, Department of History of Architecture and Urban Planning, Delft University of Technology, The Netherlands
- Yan Jia, PhD student, Tsinghua University, China
- Martin de Jong, Professor, Urban Planning, Infrastructure Development and Public Policy, Delft University of Technology, The Netherlands; Erasmus University Rotterdam, The Netherlands; Fudan University, China
- Otto Kroesen, Assistant Professor, Cross-Cultural Entrepreneurship, Delft Centre for Entrepreneurship, Delft University of Technology, The Netherlands
- Jannis Langer, MSc and PhD student, Sustainable Energy Technology, Faculty of Technology, Policy and Management, Delft University of Technology, The Netherlands
- Zhijie Li, Visiting scholar, Delft University of Technology, The Netherlands; PhD student, Harbin Institute of Technology, China; Postdoc, Leiden University, The Netherlands
- Haiyan Lu, PhD, Faculty of Technology, Policy and Management, Delft University of Technology, The Netherlands; Postdoc, Freie Universität Berlin, Germany; Peking University, China
- Dimosthenis Michalakeas, MSc, Delft University of Technology, The Netherlands; Consultant, GAIAOSE SA, Greece
- Alexander Peters, MSc, Delft University of Technology, The Netherlands

- Rahul Prasad, MSc, Delft University of Technology, The Netherlands; PhD student, Nijmegen School of Management, Radboud University, The Netherlands
- Wim Ravesteijn, Associate Professor, Sustainable and Responsible Innovation, Delft University of Technology, The Netherlands; Honorary Professor, Beijing Information Science and Technology University, China; Guest Professor, Harbin Institute of Technology, China
- Lili Song, PhD student, School of Management, Harbin Normal University, China
- Praveen Sridharan, MSc, Faculty of Technology, Policy and Management, Delft University of Technology, The Netherlands
- Frank Vollering, MSc, Faculty of Technology, Policy and Management, Delft University of Technology, The Netherlands
- Jinxin Zhang, Visiting scholar, Delft University of Technology, The Netherlands; PhD student, College of Water Conservancy and Hydropower Engineering, Hohai University, China
- Yan Zhang, Visiting scholar, Delft University of Technology, The Netherlands; Lecturer, Harbin Normal University, China
- Penglin Zhu, PhD student, Department of History of Architecture and Urban Planning, Delft University of Technology, The Netherlands

Index

Page locators in **bold** refer to tables; those in *italics* refer to figures.